Lecture Notes in Mathematics

An informal series of special lectures, seminars and reports on mathematical topics

Edited by A. Dold, Heidelberg and B. Eckmann, Zürich

16

J. Pfanzagl · W. Pierlo

Department of Mathematics
University of Cologne

Compact Systems of Sets

1966

Springer-Verlag · Berlin · Heidelberg · New York

Library of Congress Catalog Card Number 66-20636. Printed in Germany. Title No. 7336.

TABLE OF CONTENTS

Introduction

Probability measures on abstract measurable spaces have some unpleasant properties such as exhibited by examples in chapter 6, p.30, (existence of product measures) and chapter 7, p.35, (existence of regular conditional probability measures). In order to exclude such irregularities, additional assumptions have to be made. A natural, though not completely successful way is, to use topological concepts in the formulation of 'regularity conditions'.

The theory of compact systems of sets, mainly due to M a r c z e w s k i , tries to isolate those 'topologically inspired' regularity conditions which are required in order to avoid disturbing irregularities.

The purpose of the present lecture consists in a systematic presentation of M a r c z e w s k i's theory and its application to the existence of product measures, due to M a r c z e w s k i and R y l l - N a r d z e w s k i , together with its application to the existence of regular conditional probability measures, due to J i ř i n a .

The terminology is basically that of H a l m o s . A brief survey of the most important notations is to be found on page 45.

1. Compact Systems of Sets

Definition (1.1) A system $\mathcal{C} \subset \mathfrak{P}(X)$ is compact, iff for each sequence $(C_n)_{n=1,2,\ldots}$, $C_n \in \mathcal{C}$ with $\bigcap_1^\infty C_n = \emptyset$ there exists a natural number N such that $\bigcap_1^N C_n = \emptyset$.

We remark that $\mathcal{C}$ is compact iff any sequence in $\mathcal{C}$ which has the finite intersection property has a non-empty intersection. A $\cap$-system is compact iff each non-increasing sequence of non-empty sets has a non-empty intersection.

Example: Let $(X,\mathfrak{U})$ be a topological Hausdorff space and $\mathcal{C}$ the system of all compact subsets of X. Then $\mathcal{C}$ is a compact system in the sense of (1.1).

Lemma (1.2) A subsystem of a compact system is compact.

Lemma (1.3) Compactness of $\mathcal{C}$ implies compactness of $\mathcal{C}^\delta$.

Proof: Let $(K_m)_{m=1,2,\ldots}$ be a sequence in $\mathcal{C}^\delta$ with $\bigcap_1^\infty K_m = \emptyset$. From $K_m \in \mathcal{C}^\delta$ we have $K_m = \bigcap_{n=1}^\infty C_{mn}$, $C_{mn} \in \mathcal{C}$. Thus $\bigcap_{m=1}^\infty \bigcap_{n=1}^\infty C_{mn} = \emptyset$. As $\mathcal{C}$ is compact, there exists a finite subset π_0 of $\{(m,n): m,n = 1,2,\ldots\}$ such that $\bigcap_{(m,n)\in\pi_0} C_{mn} = \emptyset$.
Let M be the largest m occurring in π_0. Then $\bigcap_1^M K_m \subset \bigcap_{(m,n)\in\pi_0} C_{mn} = \emptyset$.

Lemma (1.4) Compactness of $\mathcal{C}$ implies compactness of $\mathcal{C}^\cup$.

Proof: Let $(K_m)_{m=1,2,\ldots}$ be a sequence in $\mathcal{C}^{\cup}$ with $\bigcap_1^M K_m \neq \emptyset$ for all $M = 1,2,\ldots$. We will show that $\bigcap_1^\infty K_m \neq \emptyset$.
From $K_m \in \mathcal{C}^{\cup}$, we have $K_m = \bigcup_{n=1}^{N_m} C_{mn}$, $C_{mn} \in \mathcal{C}$. Then

$$\bigcap_1^M K_m = \bigcap_{m=1}^M \bigcup_{n=1}^{N_m} C_{mn} = \bigcup \bigcap_{m=1}^M C_{mn_m},$$

where the union extends over all $(n_1,\ldots,n_M)$ with $n_m \in \{1,\ldots,N_m\}$ for $m = 1,\ldots,M$.

Because $\bigcap_1^M K_m \neq \emptyset$ for all $M = 1,2,\ldots$, to each M there exists a sequence $(n_1^{(M)},\ldots,n_M^{(M)})$ with $n_m^{(M)} \in \{1,\ldots,N_m\}$ for $m = 1,\ldots,M$ such that $\bigcap_{m=1}^M C_{mn_m^{(M)}} \neq \emptyset$. Because $n_1^{(M)} \in \{1,\ldots,N_1\}$ for all $M = 1,2\ldots$ at least one of the numbers $1,\ldots,N_1$, say n_1, occurs infinitely often. The same argument can be applied to the infinite number of sequences starting with n_1, leading to a number $n_2 \in \{1,\ldots,N_2\}$ occurring infinitely often and so on. Thus we obtain a sequence $n_1, n_2,\ldots$ such that for each $M = 1,2,\ldots$ we have $\bigcap_{m=1}^M C_{mn_m} \neq \emptyset$ (because there exist infinitely many numbers $M' \geq M$, such that $n_m^{(M')} = n_m$ for $m = 1,\ldots,M$ and $\bigcap_{m=1}^{M'} C_{mn_m^{(M')}} \neq \emptyset$). As $\mathcal{C}$ is compact, this implies $\bigcap_{m=1}^\infty C_{mn_m} \neq \emptyset$.
As $K_m \supset C_{mn_m}$, this implies: $\bigcap_1^\infty K_m \neq \emptyset$, q.e.d.

Definition (1.5) An arbitrary family of systems of subsets of X, $(\mathcal{S}_i)_{i\in I}$, is called algebraically independent resp. algebraically σ-independent, iff for each finite resp. countable (non-empty) subset $I_0 \subset I$ and each sequence $(S_i)_{i\in I_0}$ with $\emptyset \neq S_i \in \mathcal{S}_i$ for all $i \in I_0$ we have : $\bigcap_{I_0} S_i \neq \emptyset$.

<u>Theorem (1.6)</u> Let $(\mathcal{C}_i)_{i\in I}$ be a family of alg. σ-independent compact δ-systems. Then $\bigcup_I \mathcal{C}_i$ is compact.

<u>Proof:</u> Let $(C_n)_{n=1,2,\ldots}$ be a sequence in $\bigcup_I \mathcal{C}_i$ with $\bigcap_1^\infty C_n = \emptyset$. To each C_n we choose an index $i_n \in I$ such that $C_n \in \mathcal{C}_{i_n}$. Let $I_1 := \{i_n : n = 1,2,\ldots\}$. For each $i \in I_1$ we define $\pi_i := \{n : C_n \in \mathcal{C}_i\}$. Then $\bigcap_1^\infty C_n = \bigcap_{i\in I_1} \bigcap_{n\in\pi_i} C_n = \emptyset$. As I_1 is countable (and not empty), and as $\bigcap_{\pi_i} C_n \in \mathcal{C}_i$ for all $i \in I_1$, alg. σ-independence of $(\mathcal{C}_i)_{i\in I}$ implies that there exists $i_o \in I_1$ such that $\bigcap_{\pi_{i_o}} C_n = \emptyset$. Compactness of $\mathcal{C}_{i_o}$ implies that there exists a finite subset of π_{i_o}, say $\pi_{i_o} \cap \{1,\ldots,N\}$, such that $\bigcap_{\pi_{i_o}\cap\{1,\ldots,N\}} C_n = \emptyset$. Hence $\bigcap_1^N C_n = \emptyset$.

<u>Lemma (1.7)</u> Let $T: X \to Y$ be a map from X onto Y. If $\mathcal{K}$ is a compact system in Y, then $T^{-1}\mathcal{K}$ is a compact system in X.

<u>Proof:</u> Assume that $\bigcap_1^\infty T^{-1}K_n = \emptyset$. As $\bigcap_1^\infty T^{-1}K_n = T^{-1}\bigcap_1^\infty K_n$ it follows: $T^{-1}\bigcap_1^\infty K_n = \emptyset$. Then, $\bigcap_1^\infty K_n = \emptyset$ (because T maps onto Y). Compactness of $\mathcal{K}$ implies that there exists a natural number N such that $\bigcap_1^N K_n = \emptyset$. Hence $\bigcap_1^N T^{-1}K_n = T^{-1}\bigcap_1^N K_n = \emptyset$.

2. Approximation

Let $P/\mathcal{A}$ be a p-content and let $P^*/\mathfrak{P}(X)$ be the outer content induced by $P/\mathcal{A}$, i.e.

$P^*(B) = \inf\{P(A): B \subset A \in \mathcal{A}\}$ for each $B \in \mathfrak{P}(X)$

Definition (2.1) A system $\mathcal{S}$ $P/\mathcal{A}$ - approximates a set B, iff $\inf\{P^*(B - S): B \supset S \in \mathcal{S}\} = 0$.

In the following we will often make use of the following criterion:

Criterion(2.2): A system $\mathcal{S}$ $P/\mathcal{A}$ - approximates a set B iff to each $\varepsilon > 0$ there exist sets $S_\varepsilon \in \mathcal{S}$ and $A_\varepsilon \in \mathcal{A}$ such that $S_\varepsilon \subset B \subset S_\varepsilon \cup A_\varepsilon$ and $P(A_\varepsilon) < \varepsilon$.

Remarks: (i) If $B \in \mathcal{A}$, $\mathcal{S}$ $P/\mathcal{A}$ - approximates B iff to each $\varepsilon > 0$ there exist sets $S_\varepsilon \in \mathcal{S}$ and $A_\varepsilon \in \mathcal{A}$ such that $A_\varepsilon \subset S_\varepsilon \subset B$ and $P(B - A_\varepsilon) < \varepsilon$.

(ii) If $B \in \mathcal{A}, \mathcal{S} \subset \mathcal{A}$, $\mathcal{S}$ $P/\mathcal{A}$ - approximates B iff $P(B) = \sup\{P(S): B \supset S \in \mathcal{S}\}$.

(iii) Assume that $\mathcal{S}$ $P/\mathcal{A}$ - approximates B. Let $\mathcal{S} \subset \mathcal{S}'$ and $\mathcal{A} \subset \mathcal{A}'$. Then $\mathcal{S}'$ $P/\mathcal{A}'$ - approximates B.

Lemma (2.3) If $\mathcal{S}$ $P/\mathcal{A}$ - approximates the set B, there exists a set $S_0 \in \mathcal{S}^\sigma$ such that $S_0 \subset B$ and $P^*(B - S_0) = 0$.

Proof: To each n = 1,2,... there exist (according to (2.2)) sets $S_n \in \mathcal{S}$ and $A_n \in \mathcal{A}$ such that

$$S_n \subset B \subset S_n \cup A_n \quad \text{and} \quad P(A_n) < \frac{1}{n}.$$

Let $S_0 := \bigcup_1^\infty S_n$. Then $S_0 \subset B$. Furthermore, $B - S_n \subset A_n$. Hence $B - S_0 \subset \bigcap_1^\infty A_n$. As $P^*(\bigcap_1^\infty A_n) \leq P(A_m) < \frac{1}{m}$ for each $m = 1,2,\ldots$, we have $P^*(\bigcap_1^\infty A_n) = 0$. Thus $P^*(B - S_0) = 0$.

Corollary (2.4) If $B \in \mathcal{A}$, $\mathcal{S} \subset \mathcal{A}$ and $\mathcal{A}$ is a σ-algebra, we have $P(S_0) = P(B)$. (As $B - S_0 \in \mathcal{A}$ and $S_0 \subset B$, we have $P(B) = P(S_0) + P(B - S_0)$. By (2.3) $P(B - S_0) = P^*(B - S_0) = 0$.)

Definition (2.5) A system $\mathcal{S}$ $P/\mathcal{A}$ - approximates a system $\mathcal{B}$, iff $\mathcal{S}$ $P/\mathcal{A}$ - approximates each $B \in \mathcal{B}$.

Lemma (2.6)

(a) If $\mathcal{S}$ $P/\mathcal{A}$ - approximates $\mathcal{B}$, then $\mathcal{S}^{\cup}$ $P/\mathcal{A}$ - approximates $\mathcal{B}^{\cup}$, $\mathcal{S}^{\cap}$ $P/\mathcal{A}$ - approximates $\mathcal{B}^{\cap}$.

(b) If $P/\mathcal{A}$ is a p-measure, then the assertions of (a) even hold for countable unions and intersections.

Proof: Let O stand for any of the operations $\cup$ or $\cap$. If $B \in \mathcal{B}^O$, there exist sets $B_1,\ldots,B_m$ with $B_i \in \mathcal{B}$, $i = 1,\ldots,m$, such that $B = \mathop{O}_1^m B_i$. As $\mathcal{S}$ $P/\mathcal{A}$ - approximates $\mathcal{B}$, to each $\varepsilon > 0$ and each i there exist sets $S_i \in \mathcal{S}$ and $A_i \in \mathcal{A}$ such that $S_i \subset B_i \subset S_i \cup A_i$ and $P(A_i) < \frac{\varepsilon}{2^i}$. Therefore $\mathop{O}_1^m S_i \subset \mathop{O}_1^m B_i \subset \mathop{O}_1^m (S_i \cup A_i) \subset (\mathop{O}_1^m S_i) \cup (\bigcup_1^m A_i)$ and $P(\bigcup_1^m A_i) \leq \sum_1^m P(A_i) < \varepsilon$.
If $P/\mathcal{A}$ is a p-measure, this proof works even for the case of countable unions and intersections.

Lemma (2.7) If $\mathcal{T}$ $P/\mathcal{A}$ - approximates $\mathcal{S}$ and $\mathcal{S}$ $P/\mathcal{A}$ - approximates $\mathcal{B}$, then $\mathcal{T}$ $P/\mathcal{A}$ - approximates $\mathcal{B}$.

Proof: Let $B \in \mathcal{B}$. To each $\varepsilon > 0$ there exist sets $S_\varepsilon \in \mathcal{S}$ and $A_\varepsilon \in \mathcal{A}$ such that $S_\varepsilon \subset B \subset S_\varepsilon \cup A_\varepsilon$ and $P(A_\varepsilon) < \frac{\varepsilon}{2}$. Furthermore, to S_ε there exist sets $T_\varepsilon \in \mathcal{T}$ and $A'_\varepsilon \in \mathcal{A}$ such that $T_\varepsilon \subset S_\varepsilon \subset T_\varepsilon \cup A'_\varepsilon$ and $P(A'_\varepsilon) < \frac{\varepsilon}{2}$. Therefore $T_\varepsilon \subset B \subset T_\varepsilon \cup A'_\varepsilon \cup A_\varepsilon$ and $P(A'_\varepsilon \cup A_\varepsilon) \leq P(A'_\varepsilon) + P(A_\varepsilon) < \varepsilon$.

Lemma (2.8) Let $P/\mathcal{A}^*$ be a p-measure. If $\mathcal{S}$ $P/\mathcal{A}^*$ - approximates $\mathcal{A}^*$, then also $\mathcal{S}^\delta \cap \mathcal{A}^*$ $P/\mathcal{A}^*$ - approximates $\mathcal{A}^*$.

Proof: Let $A \in \mathcal{A}^*$. We define a sequence $A_1, A_2, \ldots$ by induction: $A_1 := A$. If $A_n \in \mathcal{A}^*$ is defined, we choose $S_n \in \mathcal{S}$ and $B_n \in \mathcal{A}^*$ such that $S_n \subset A_n \subset S_n \cup B_n$ and $P(B_n) < \frac{\varepsilon}{2^n}$. From $A_n \subset S_n \cup B_n$ we obtain $A_n - B_n \subset S_n$. Now we define:

$$(+) \qquad A_{n+1} := A_n - B_n,$$

$A_{n+1} \in \mathcal{A}^*$. Thus we obtain sequences $\{A_n\} \subset \mathcal{A}^*$, $\{B_n\} \subset \mathcal{A}^*$ and $\{S_n\} \subset \mathcal{S}$ such that $A_{n+1} \subset S_n \subset A_n$. Hence

$$(++) \qquad \bigcap_1^\infty S_n = \bigcap_1^\infty A_n \in \mathcal{S}^\delta \cap \mathcal{A}^*.$$

From (+) we have : $A_n \subset A_{n+1} \cup B_n$. Hence $A_1 \subset A_n \cup (\bigcup_1^{n-1} B_k) \subset A_n \cup (\bigcup_1^\infty B_k)$ for all $n = 1,2,\ldots$. Therefore $A_1 \subset (\bigcap_1^\infty A_n) \cup (\bigcup_1^\infty B_k)$. Together with (++), this implies: $\bigcap_1^\infty S_n \subset A_1 \subset (\bigcap_1^\infty S_n) \cup (\bigcup_1^\infty B_k)$, where $P(\bigcup_1^\infty B_k) \leq \sum_1^\infty P(B_k) < \varepsilon$.

Lemma (2.9) If $\mathcal{A}$ is an algebra and $P/\sigma(\mathcal{A})$ is a p-measure, then $\mathcal{A}^\delta$ $P/\sigma(\mathcal{A})$ - approximates the σ-algebra $\sigma(\mathcal{A})$.

Proof: From the extension theorem (see Halmos, Theorem A, p.54) we have: $P(A) = \inf\{\sum_1^\infty P(A_i): A \subset \bigcup_1^\infty A_i, A_i \in \mathcal{A}, i = 1,2,\ldots\}$ for each $A \in \sigma(\mathcal{A})$. From this it follows, that $P(A) = \inf\{P(B): A \subset B, B \in \mathcal{A}^\sigma\}$ for each $A \in \sigma(\mathcal{A})$ which implies $P(A) = \sup\{P(B): B \subset A, B \in \mathcal{A}^\delta\}$.

Theorem (2.1o) Let $\mathcal{A}$ be a sub-algebra of a σ-algebra $\mathcal{A}^*$ and $P/\mathcal{A}^*$ be a p-measure. If $\mathcal{S}$ $P/\mathcal{A}^*$ - approximates $\mathcal{A}$, then $\mathcal{S}^\delta$ $P/\mathcal{A}^*$ - approximates $\sigma(\mathcal{A})$.

Proof: According to (2.6 b) $\mathcal{S}^\delta$ $P/\mathcal{A}^*$ - approximates $\mathcal{A}^\delta$, according to (2.9) $\mathcal{A}^\delta$ $P/\mathcal{A}^*$ - approximates $\sigma(\mathcal{A})$. Therefore $\mathcal{S}^\delta$ $P/\mathcal{A}^*$ - approximates $\sigma(\mathcal{A})$ according to (2.7).

Remark: If in (2.10) $\mathcal{A}^* = \sigma(\mathcal{A})$, we have from (2.8) that already $\mathcal{S}^\delta \cap \mathcal{A}^*$ $P/\mathcal{A}^*$ - approximates $\mathcal{A}^*$.

Lemma (2.11) Let $P/\mathcal{A}$ be a p-content and I an arbitrary index set. Assume that for each $i \in I$, $\mathcal{S}_i$ $P/\mathcal{A}$ - approximates $\mathcal{B}_i$ (with $\mathcal{S}_i, \mathcal{B}_i \subset \mathfrak{P}(X)$). Then $\bigcup_I \mathcal{S}_i$ $P/\mathcal{A}$ - approximates $\bigcup_I \mathcal{B}_i$.

Theorem (2.12) Let $P/\mathcal{A}$ be a p-content. Assume that for each i of an arbitrary index set I $\mathcal{S}_i$ $P/\mathcal{A}$ - approximates the algebra $\mathcal{B}_i$. Then $(\bigcup_I \mathcal{S}_i)^{\cap\cup}$ $P/\mathcal{A}$ - approximates $\alpha(\bigcup_I \mathcal{B}_i)$.

Proof: Let $B \in \alpha(\bigcup_I \mathcal{B}_i)$. Then, by (0.2) there exists a finite set $I_o \subset I$ such that $B \in (\bigcup_{I_o} \mathcal{B}_i)^{\cap\cup}$. According to (2.11) and (2.6 a) $(\bigcup_{I_o} \mathcal{S}_i)^{\cap\cup}$ $P/\mathcal{A}$ - approximates $(\bigcup_{I_o} \mathcal{B}_i)^{\cap\cup}$ and therefore it approximates the set B. As $(\bigcup_{I_o} \mathcal{S}_i)^{\cap\cup} \subset (\bigcup_I \mathcal{S}_i)^{\cap\cup}$, also $(\bigcup_I \mathcal{S}_i)^{\cap\cup}$ $P/\mathcal{A}$ - approximates B.

Lemma (2.13) Let $(X, \mathcal{A}^*)$ and $(Y, \mathcal{B}^*)$ be two measurable spaces and $P/\mathcal{A}^*$ a p-measure. Let $T: X \to Y$ be an $\mathcal{A}^*, \mathcal{B}^*$ - measurable map from X into Y. Let $P_T/\mathcal{B}^*$ be the p-measure induced by T and $P/\mathcal{A}^*$ (i.e. $P_T(D) := P(T^{-1}D)$). If $\mathcal{E}$ $P_T/\mathcal{B}^*$ - approximates $\mathcal{F}$ ($\mathcal{E}, \mathcal{F} \subset \mathfrak{P}(Y)$), then $T^{-1}\mathcal{E}$ $P/T^{-1}\mathcal{B}^*$ - approximates $T^{-1}\mathcal{F}$.

__Proof:__ Let $\varepsilon > 0$ and consider $T^{-1}F$ with $F \in \mathfrak{F}$. There exist two sets $E_\varepsilon \in \mathcal{E}$ and $D_\varepsilon \in \mathfrak{D}^*$ such that

$$E_\varepsilon \subset F \subset E_\varepsilon \cup D_\varepsilon \quad \text{and} \quad P_T(D_\varepsilon) < \varepsilon .$$

Then, $T^{-1}E_\varepsilon \subset T^{-1}F \subset T^{-1}E_\varepsilon \cup T^{-1}D_\varepsilon$ and $P(T^{-1}D_\varepsilon) = P_T(D_\varepsilon) < \varepsilon$.

3. Compact Approximation

Let $P/\mathcal{A}$ be a p-content.

Definition (3.1) A set B is compact $P/\mathcal{A}$ - approximable, iff there exists a compact system of sets which $P/\mathcal{A}$ - approximates B.

Definition (3.2) A system of sets is compact $P/\mathcal{A}$ - approximable iff each set is compact $P/\mathcal{A}$ - approximable by the same compact system.

Lemma (3.3) Let the system $\mathcal{B} \subset \mathcal{A}$ be compact $P/\mathcal{A}$ - approximable. Then for each non-increasing sequence $(B_n)_{n=1,2...}$ with $B_n \in \mathcal{B}$ and $\bigcap_1^\infty B_n = \emptyset$, we have $P(B_n) \downarrow 0$ for $n \to \infty$.

Proof: Let $\mathcal{C}$ be the compact system $P/\mathcal{A}$ - approximating $\mathcal{B}$. To each $\varepsilon > 0$ and each B_n there exist sets $C_{n,\varepsilon} \in \mathcal{C}$ and $A_{n,\varepsilon} \in \mathcal{A}$ such that

$$C_{n,\varepsilon} \subset B_n \subset C_{n,\varepsilon} \cup A_{n,\varepsilon} \quad \text{and} \quad P(A_{n,\varepsilon}) < \frac{\varepsilon}{2^n}.$$

As $\bigcap_1^\infty B_n = \emptyset$, we have $\bigcap_1^\infty C_{n,\varepsilon} = \emptyset$. Thus, there exists N_ε such that $\bigcap_1^{N_\varepsilon} C_{n,\varepsilon} = \emptyset$. As $B_n \subset B_k \subset C_{k,\varepsilon} \cup A_{k,\varepsilon}$ for any $k = 1,2,\dots,n$, we have $B_n \subset (\bigcap_1^n C_{k,\varepsilon}) \cup (\bigcup_1^n A_{k,\varepsilon})$. Therefore, for $n \geq N_\varepsilon$ we have $B_n \subset \bigcup_1^n A_{k,\varepsilon}$ and $P(B_n) \leq P(\bigcup_1^n A_{k,\varepsilon}) \leq \sum_1^n P(A_{k,\varepsilon}) < \varepsilon$.

Theorem (3.4) Let the semi-algebra $\mathcal{B}_0 \subset \mathcal{A}$ be compact $P/\mathcal{A}$ - approximable. Then $P/\mathcal{B}_0$ is σ-additiv and there exists a (uniquely determined) extension to a measure $P/\sigma(\mathcal{B}_0)$. +)

+) As the extension of $P/\mathcal{B}_0$ to a measure on $\sigma(\mathcal{B}_0)$ is uniquely determined, we will denote this measure with P too, if no ambiguity is possible.

Proof: As $\mathcal{B}_0$ is $P/\mathcal{A}$ - approximable by a compact system $\mathcal{C}$, the algebra $\mathcal{B}_0^+$ (the smallest algebra containing $\mathcal{B}_0$) is $P/\mathcal{A}$ - approximable by the system $\mathcal{C}^{\cup}$, which is compact again. According to (3.3), for any non-increasing sequence $(B_n)_{n=1,2,\ldots}$ with $B_n \in \mathcal{B}_0^+$ and $\bigcap_1^\infty B_n = \emptyset$, we have $P(B_n) \downarrow 0$ (for $n \to \infty$). This implies σ-additivity of $P/\mathcal{B}_0^+$. Hence P is also σ-additive on $\mathcal{B}_0$ and according to the extension theorem, there exists an (uniquely determined) extension of $P/\mathcal{B}_0$ to a measure $P/\sigma(\mathcal{B}_0)$.

Corollary (3.5) If the algebra $\mathcal{A}$ is compact $P/\mathcal{A}$ - approximable, then $\sigma(\mathcal{A})$ is compact $P/\sigma(\mathcal{A})$ - approximable (where $P/\sigma(\mathcal{A})$ is the (uniquely determined) extension of $P/\mathcal{A}$ to a measure on $\sigma(\mathcal{A})$).

Proof: That $P/\mathcal{A}$ has an (uniquely determined) extension to a measure $P/\sigma(\mathcal{A})$ follows from (3.4).

Let $\mathcal{C}$ be a compact system $P/\mathcal{A}$ - approximating $\mathcal{A}$. Then $\mathcal{C}$ $P/\sigma(\mathcal{A})$ - approximates $\mathcal{A}$. Therefore, according to (2.10), $\mathcal{C}^\delta$ $P/\sigma(\mathcal{A})$ - approximates $\sigma(\mathcal{A})$. According to (1.3), $\mathcal{C}^\delta$ is compact, what concludes the proof.

The following lemma shows that (under the assumptions of (3.5)) $\sigma(\mathcal{A})$ is $P/\sigma(\mathcal{A})$ - approximable even by a compact system contained in $\sigma(\mathcal{A})$.

Lemma (3.6) Let $P/\mathcal{A}^*$ be a p-measure. If $\mathcal{A}^*$ is compact $P/\mathcal{A}^*$ - approximable, then there exists a compact δ-system contained in $\mathcal{A}^*$ $P/\mathcal{A}^*$ - approximating $\mathcal{A}^*$.

Lemma (3.6) is an immediate consequence of (2.8)(together with (1.2) and(1.3)).

<u>Lemma (3.7)</u> Let $(X,\mathcal{A}^*)$ and $(Y,\mathcal{B}^*)$ be two measurable spaces and $P/\mathcal{A}^*$ be a p-measure. Let $T: X \to Y$ be a map from X into Y with the following properties

(i) T is $\mathcal{A}^*,\mathcal{B}^*$ - measurable,

(ii) T(X) is $P_T/\mathcal{B}^*$ - approximable by $\mathcal{B}^*$.

If $\mathcal{B}^*$ is compact $P_T/\mathcal{B}^*$ - approximable, then $T^{-1}\mathcal{B}^*$ is compact $P/T^{-1}\mathcal{B}^*$- approximable.

<u>Proof:</u> Let $\mathcal{A}_T^*: = \{E \subset Y: T^{-1}E \in \mathcal{A}^*\}$. Then $\mathcal{B}^* \subset \mathcal{A}_T^*$, $T(X) \in \mathcal{A}_T^*$, and from (ii) we have that T(X) is $P_T/\mathcal{A}_T^*$ - approximable by $\mathcal{B}^*$. Therefore, according to (2.4), there exists $Y_0 \in \mathcal{B}^*$ such that $Y_0 \subset T(X)$ and $P_T(Y_0) = 1$. Let $\mathcal{K}$ be a compact system $P_T/\mathcal{B}^*$ - approximating $\mathcal{B}^*$. Let $\mathcal{K}_0: = \{K \in \mathcal{K}: K \subset Y_0\}$. According to (1.2), $\mathcal{K}_0$ is compact. We will show that $\mathcal{K}_0$ $P_T/\mathcal{B}^*$ - approximates $\mathcal{B}^*$: Let $D \in \mathcal{B}^*$. Then $D \cap Y_0 \in \mathcal{B}^*$, whence for any $\varepsilon > 0$ there exist sets $K_\varepsilon \in \mathcal{K}$ and $D_\varepsilon \in \mathcal{B}^*$ such that

$$K_\varepsilon \subset D \cap Y_0 \subset K_\varepsilon \cup D_\varepsilon \quad \text{and} \quad P_T(D_\varepsilon) < \varepsilon .$$

Hence $K_\varepsilon \subset D \subset K_\varepsilon \cup D_\varepsilon \cup Y_0^c$ and $P_T(D_\varepsilon \cup Y_0^c) = P_T(D_\varepsilon) < \varepsilon$. As $K_\varepsilon \in \mathcal{K}$ and $K_\varepsilon \subset Y_0$, we have $K_\varepsilon \in \mathcal{K}_0$.

As all sets of $\mathcal{K}_0$ are contained in T(X), the system $T^{-1}\mathcal{K}_0$ is compact according to (1.7), and it $P/T^{-1}\mathcal{B}^*$ - approximates $T^{-1}\mathcal{B}^*$ according to (2.13).

4. Compact Approximation in Topological Spaces

Let $(X,\mathcal{U})$ be an arbitrary topological Hausdorff space ($\mathcal{U}$ being the class of open sets).

Definition (4.1) The Borel - algebra of $(X,\mathcal{U})$ is the smallest σ-algebra containing all open sets. It will be denoted by $\mathcal{B}^*$.+)

Remark: As each closed set is the complement of an open set, the Borel - algebra also contains all closed sets of $(X,\mathcal{U})$. As in a Hausdorff - space each compact set is closed, the Borel - algebra contains all compact sets of $(X,\mathcal{U})$.

Definition (4.2) A p-content $P/\mathcal{B}^*$ on the Borel - algebra $\mathcal{B}^*$ is regular, iff the system of compact sets, say $\mathcal{K}$, $P/\mathcal{B}^*$ - approximates $\mathcal{U}$.

Lemma (4.3) If $P/\mathcal{B}^*$ is a regular p-measure, then $\mathcal{K}$ $P/\mathcal{B}^*$ - approximates $\mathcal{B}^*$.

Proof: Let $\mathcal{A}$ be the system of all closed sets given by the topology $\mathcal{U}$. Then $\alpha(\mathcal{U}) = (\mathcal{A} \cup \mathcal{U})^{\cap\cup}$. By definition of regularity (4.2) the compact system $\mathcal{K}$ $P/\mathcal{B}^*$ - approximates $\mathcal{U}$. If we show that $\mathcal{K}$ also $P/\mathcal{B}^*$ - approximates $\mathcal{A}$, it follows from (2.6 a) that $\mathcal{K}^{\cap\cup} = \mathcal{K}$ $P/\mathcal{B}^*$ - approximates $\alpha(\mathcal{U})$. Hence - as $P/\mathcal{B}^*$ is a measure - according to (2.10) $\mathcal{K}^{\delta} = \mathcal{K}$ $P/\mathcal{B}^*$ - approximates $\sigma(\alpha(\mathcal{U})) = \sigma(\mathcal{U}) = \mathcal{B}^*$.

+) For this definition see e.g. H e w i t t and R o s s , p.118. The Borel algebra defined here is greater than the Borel - algebra defined by H a l m o s (H a l m o s , p.219).

It remains to show that $\mathcal{K}$ $P/\mathcal{B}^*$ - approximates $\mathcal{A}$: As X is open, to each $\varepsilon > 0$ there exists a set $K_\varepsilon \in \mathcal{K}$ such that $P(X - K_\varepsilon) < \varepsilon$. If A is closed, then $A \cap K_\varepsilon$ is compact ($A \cap K_\varepsilon \in \mathcal{K}$), $A \cap K_\varepsilon \subset A$ and $P(A - A \cap K_\varepsilon) \leq P(X - K_\varepsilon) < \varepsilon$.

Definition (4.4) A Hausdorff topology is tight, iff each open set is σ-compact (i.e. countable union of compact sets).

Remark: In a tight Hausdorff topology each p-measure $P/\mathcal{B}^*$ is regular.

Criterion (4.5) A Hausdorff topology is tight iff X is σ-compact and each open set is F_σ (or each closed set is G_δ).

Proof: Necessity. If a topological Hausdorff space is tight, each open set is the countable union of compact sets. As each compact set is closed, each open set is F_σ. As X itself is open, it is σ-compact.

Sufficiency. If in a σ-compact topological Hausdorff space each open set is a F_σ, we have $X = \bigcup_1^\infty K_n$, where K_n is compact for n = 1,2,... and $U = \bigcup_1^\infty A_m$, where A_m is closed for m = 1,2..., for each open set U. Therefore $U = \bigcup_{m=1}^\infty \bigcup_{n=1}^\infty A_m \cap K_n$, i.e., U is the union of the countable number of sets $A_m \cap K_n$ which are closed subsets of compact sets and hence compact.

It remains to show that there are Hausdorff topologies in which each open set is a F_σ.

Lemma (4.6)[+)] In a metric space, each open set is F_σ.

+) See S i e r p i n s k i , p.166.

Proof: We prove the equivalent statement that each closed set is G_δ. Let A be an arbitrary closed set. Let ρ be the distance function of the given metric space.
Let $B(x_o,r) := \{x: \rho(x_o,x) < r\}$ for any $x_o \in X$ and any $r \in R^1$.
We define $U_n := \bigcup_{x \in A} B(x,\frac{1}{n})$ for $n = 1,2,\ldots$. U_n is open and from $A \subset U_n$ for $n = 1,2,\ldots$ we obtain $A \subset \bigcap_1^\infty U_n$. We show: $A = \bigcap_1^\infty U_n$. Let $x \in \bigcap_1^\infty U_n$. Then $x \in U_n$ for each n and therefore for each $n = 1,2,\ldots$ there exists $x_n \in A$ such that $x \in B(x_n,\frac{1}{n})$. Hence $\rho(x_n,x) < \frac{1}{n}$ for $n = 1,2,\ldots$, which implies $x_n \to x$ (for $n \to \infty$). As $x_n \in A$ and A is closed, we have $x \in A$.

For the following lemma we need the notions 'regular-', and 'T_1-space': A topological space is regular iff each neighbourhood contains a closed neighbourhood. A topological space is a T_1-space iff for each $x \in X$ the set $\{x\}$ is closed.

Lemma (4.7) In a regular (Hausdorff) topology with countable base, each open set is F_σ.

Proof: A direct proof can be given as follows: Let U be an arbitrary open set. By definition of regularity, to each $x \in U$ there exists a closed set A_x such that $x \in A_x^{(i)} \subset A_x \subset U$ (where $A_x^{(i)}$ is the (non-empty) interior of A_x). The system $\{A_x^{(i)}: x \in U\}$ is an open cover of U. As the topological space has a countable base, according to Lindelöfs theorem (K e l l e y , 15 Theorem, p.49) this cover contains a countable subcover, say $(A_{x_n}^{(i)})_{n=1,2,\ldots}$. Because $U \subset \bigcup_1^\infty A_{x_n}^{(i)} \subset \bigcup_1^\infty A_{x_n} \subset U$, U is F_σ.

We remark that in case of a regular T_1-space with countable base, (4.7) follows immediately from the fact that each regular topological T_1-space with countable base is metrizable (see K e l l e y , 17 Theorem, p.125).

<u>Lemma (4.8)</u> If a Hausdorff topology is tight, then for each subbase ω of $\mathcal{U}$ we have $\mathcal{U} \subset \omega^{\cap\sigma}$ and therefore $\sigma(\omega) = \sigma(\mathcal{U})$.

<u>Proof:</u> Let $U \in \mathcal{U}$. As $\mathcal{U}$ is tight, $U = \bigcup_1^\infty K_n$, $K_n \in \mathcal{K}$. To each $x \in K_n$ there exists $V_x \in \omega^\cap$ such that $x \in V_x \subset U$. The system $\{V_x: x \in K_n\}$ is an open cover of K_n. As K_n is compact, the cover contains a finite subcover, say $V_1^{(n)}, \dots, V_{m_n}^{(n)}$: $K_n \subset \bigcup_{i=1}^{m_n} V_i^{(n)} \subset U$. Hence, $U = \bigcup_1^\infty K_n \subset \bigcup_{n=1}^\infty \bigcup_{i=1}^{m_n} V_i^{(n)} \subset U$. Therefore, $U \in \omega^{\cap\sigma}$.

We remark, that in a locally compact Hausdorff topology the condition $\mathcal{U} \subset \omega^{\cap\sigma}$ for each subbase ω is also sufficient for $\mathcal{U}$ being tight.

Let I be an arbitrary index set, $(\mathcal{U}_i)_{i \in I}$ a family of Hausdorff topologies over the same space X and $(\mathcal{B}_i^*)_{i \in I}$ the corresponding family of Borel - algebras.

The product topology $\bigtimes_I \mathcal{U}_i$ is the smallest topology containing $\bigcup_I \mathcal{U}_i$.+) Let $\mathcal{B}^*$ be the Borel - algebra corresponding to the product topology.

+) This definition includes the usual one, which defines the product topology of topological spaces $(X_i, \mathcal{U}_i')$, $i \in I$, on the cartesian product $X := \bigtimes_I X_i$ (see K e l l e y , pp.88-90): Putting $\mathcal{U}_i = Z_{\{i\}}(\mathcal{U}_i')$ (for the notation see p.17) the product topology $\bigtimes_I \mathcal{U}_i$ in our sense is the same as the product topology of $\mathcal{U}_i'$, $i \in I$, in the usual sense.

The product-σ-algebra, say $\underset{I}{\times}\mathcal{B}_i^*$, is the smallest σ-algebra containing $\underset{I}{\cup}\mathcal{B}_i^*$.

We have:

$$\mathcal{U}_i \subset \underset{I}{\cup}\, \mathcal{U}_i \subset \underset{I}{\times}\mathcal{U}_i \quad \text{for each } i \in I.$$

Hence,

$$\mathcal{B}_i^* = \sigma(\mathcal{U}_i) \subset \sigma(\underset{I}{\times}\mathcal{U}_i) = \mathcal{B}^* \text{ for each } i \in I.$$

Therefore,

$$\underset{I}{\times}\mathcal{B}_i^* \subset \mathcal{B}^*.$$

In general the product-σ-algebra of Borel - algebras is <u>not</u> equal to the Borel - algebra corresponding to the product topology as is shown by the following example:

<u>Example:</u> Let $I = R^1$, $X_i = \{0,1\}$ for $i \in I$ and $X = \underset{I}{\times} X_i$. Let $\mathcal{B} = \mathcal{U} = \{\{0\}, \{1\}, \{0,1\}, \emptyset\}$ and $\mathcal{B}_i^* = Z_{\{i\}}(\mathcal{B})$, $\mathcal{U}_i = Z_{\{i\}}(\mathcal{U})$ for each $i \in I$ (where $Z_{\{i\}}(\mathcal{B}) := \{Z_{\{i\}}(B) : B \in \mathcal{B}\}$ is the cylinder-σ-algebra of $\mathcal{B}$ on X and $Z_{\{i\}}(B) := \{x \in X : x_i \in B\}$ is the cylinder of B over the components X_j with $j \in I - \{i\}$, and $Z_{\{i\}}(\mathcal{U})$ resp. $Z_{\{i\}}(U)$ are defined in the same manner). Then the product-σ-algebra $\underset{I}{\times}\mathcal{B}_i^*$ is not equal to the Borel - algebra $\mathcal{B}^*$ corresponding to the product topology $\underset{I}{\times}\mathcal{U}_i$.

<u>Proof:</u> For each $B \in \underset{I}{\times}\mathcal{B}_i^*$ there exists a countable subset $I_o \subset I$ and a set $B_{I_o} \subset \underset{I_o}{\times} X_i$ such that

$$(+) \qquad B = Z_{I_o}(B_{I_o})$$

where the cylinder Z_{I_o} is to be taken over the components X_i with $i \in I - I_o$. This statement follows immediately from the fact that the system of all subsets of X which can be represented in the form (+), is a σ-algebra containing $\bigcup_I \mathcal{B}_i^*$.

Let for each $i \in I$ $B_i := \{x \in X: x_i = 0\}$. As $B_i \in \mathcal{U}_i$ we have $B^* := \bigcup_I B_i \in \bigtimes_I \mathcal{U}_i \subset \mathcal{B}^*$. We will prove that $B^* \notin \bigtimes_I \mathcal{B}_i^*$: Assume that $B^* \in \bigtimes_I \mathcal{B}_i^*$. Then there exists a countable subset $I_0 \subset I$ and a set $B_{I_0}^* \subset \bigtimes_{I_0} X_i$ such that $B^* = Z_{I_0}(B_{I_0}^*)$. Let $i \in I - I_0$. $B_i \subset B^*$ implies $B_{I_0}^* = \bigtimes_{I_0} X_i$, i.e. $B^* = X$. This is a contradiction, for the function $x \in X$ with $x_i = 1$ for each $i \in I$ does not belong to B^*.

Sufficient for the coincidence of both σ-algebras are the following conditions:

Lemma (4.9) If I is countable and each topology $\mathcal{U}_i$, $i \in I$, has a countable subbase, then $\bigtimes_I \mathcal{B}_i^* = \mathcal{B}^*$.

Proof: Let $\mathcal{S}_i$, $i \in I$, be a countable subbase for $\mathcal{U}_i$. Then, $\bigcup_I \mathcal{S}_i$ is a subbase of the product topology. As I is countable, this subbase is countable and therefore $\bigtimes_I \mathcal{U}_i = (\bigcup_I \mathcal{S}_i)^{\cap\sigma}$. Thus, $\mathcal{B}^* = \sigma(\bigcup_I \mathcal{S}_i)$. As $\mathcal{S}_i \subset \mathcal{B}_i^*$ for all $i \in I$, we have $\mathcal{B}^* \subset \sigma(\bigcup_I \mathcal{B}_i^*) = \bigtimes_I \mathcal{B}_i^*$.

Lemma (4.10) If the product topology is tight, $\bigtimes_I \mathcal{B}_i^* = \mathcal{B}^*$.

Proof: $\bigcup_I \mathcal{U}_i$ is a subbase for the product topology. As the product topology is tight, we have according to (4.8) that $\mathcal{B}^* = \sigma(\bigcup_I \mathcal{U}_i)$. As $\mathcal{U}_i \subset \mathcal{B}_i^*$ for all $i \in I$, we obtain $\mathcal{B}^* \subset \sigma(\bigcup_I \mathcal{B}_i^*) = \bigtimes_I \mathcal{B}_i^*$.

Lemma (4.11) Let I be countable and assume that for each $i \in I$ a Hausdorff topology $\mathcal{U}_i$ is given such that the family $(\mathcal{U}_i^c)_{i \in I}$ is alg. σ-independent. Let $\mathcal{K}_i, i \in I$, be the system of compact sets in the topology given by $\mathcal{U}_i$. Then the sets of the system $\mathcal{K}' := \{\bigcap_I K_i : K_i \in \mathcal{K}_i, i \in I\}$ are compact in the product topology $\bigtimes_I \mathcal{U}_i$.

Proof: Let $\mathcal{S}_i$, $i \in I$, be a subbase of $\mathcal{U}_i$. Then $\bigcup_I \mathcal{S}_i$ is a subbase for the product topology $\bigtimes_I \mathcal{U}_i$. According to a theorem by A l e x a n d e r (K e l l e y , 6 Theorem, p.139) it is sufficient to show that any cover of an arbitrary $K \in \mathcal{K}'$ by elements of the subbase $\bigcup_I \mathcal{S}_i$ contains a finite subcover. Let $\mathcal{T} \subset \bigcup_I \mathcal{S}_i$ be a cover of K. Let $\mathcal{T}_i := \mathcal{T} \cap \mathcal{S}_i$ and $T_i := \bigcup_{T \in \mathcal{T}_i} T$. Then, $K = \bigcap_I K_i \subset \bigcup_I T_i$ and therefore $\bigcap_I (K_i \cap T_i^c) = \emptyset$. As $T_i \in \mathcal{U}_i$, we have $K_i \cap T_i^c \in \mathcal{U}_i^c$. As these systems are alg. σ-independent, there exists i_o such that $K_{i_o} \cap T_{i_o}^c = \emptyset$, i.e. $K_{i_o} \subset T_{i_o}$. Hence $\mathcal{T}_{i_o}$ is a cover of K_{i_o}. As K_{i_o} is compact in $\mathcal{U}_{i_o}$, there exists a finite subsystem, say $\mathcal{T}_{i_o}'$, of $\mathcal{T}_{i_o}$ covering K_{i_o}, and hence

$$\bigcap_I K_i \subset K_{i_o} \subset \bigcup_{T \in \mathcal{T}_{i_o}'} T.$$

Lemma (4.12) Let I be countable and assume that for each $i \in I$ a Hausdorff topology $\mathcal{U}_i$ with countable subbase is given such that the family $(\mathcal{U}_i^c)_{i \in I}$ is alg. σ-independent. Let $P/\mathcal{B}^*$ be a p-measure with the property that $P/\mathcal{B}_i^*$ is regular for each $i \in I$. Then $P/\mathcal{B}^*$ is regular.

Proof: Let $\mathcal{S}_i$, $i \in I$, be a countable subbase of $\mathcal{U}_i$ and $\mathcal{K}_i$ be the system of compact sets in the topology given by $\mathcal{U}_i$. Since $\mathcal{S}_i \subset \mathcal{U}_i$, $i \in I$, we have by definition of regularity that $\mathcal{K}_i$ $P/\mathcal{B}_i^*$ - approximates $\mathcal{S}_i$. Then, $\mathcal{K}_i$ $P/\mathcal{B}^*$ - approximates $\mathcal{S}_i$.

$\bigcup_I \mathcal{S}_i$ is a countable subbase and $\mathcal{S} := (\bigcup_I \mathcal{S}_i)^\cap$ a countable base of $\bigtimes_I \mathcal{U}_i$. As we can assume without restriction of generality that $X \in \mathcal{S}_i$, $i \in I$, each set out of $\mathcal{S}$ is of the form $\bigcap_I S_{ij_i}$ with $S_{ij_i} \in \mathcal{S}_i^\cap$ and $S_{ij_i} = X$ for all but a finite number of indices $i \in I$. Since $\mathcal{K}_i^\cap = \mathcal{K}_i$ $P/\mathcal{B}^*$ - approximates $\mathcal{S}_i^\cap$, we have (see 2.6 b)

(+) $\quad \mathcal{K}' := \{\bigcap_I K_i : K_i \in \mathcal{K}_i,\ i \in I\}$ $P/\mathcal{B}^*$ - approximates $\mathcal{S}$.

By (4.11) we have $\mathcal{K}' \subset \mathcal{K}$, the system of compact sets given by the product topology $\bigtimes_I \mathcal{U}_i$. Thus we obtain from (+): $\mathcal{K}$ $P/\mathcal{B}^*$ - approximates $\mathcal{S}$. Therefore, according to (2.6 b), $\mathcal{K}^\sigma$ $P/\mathcal{B}^*$ - approximates $\mathcal{S}^\sigma = \bigtimes_I \mathcal{U}_i$. Since $\bigtimes_I \mathcal{U}_i$ is a Hausdorff topology, $\mathcal{K}$ $P/\mathcal{B}^*$ - approximates $\mathcal{K}^\sigma$, whence from (2.7) we obtain: $\mathcal{K}$ $P/\mathcal{B}^*$ - approximates $\bigtimes_I \mathcal{U}_i$, q.e.d.

Lemma (4.13) Let I be finite and $\mathcal{U}_i$, $i \in I$, be a Hausdorff topology such that the family $(\mathcal{U}_i^c)_{i\in I}$ is alg. independent. Then, if each $\mathcal{U}_i$, $i \in I$, is tight with countable subbase, $\bigtimes_I \mathcal{U}_i$ is tight (with countable subbase).

Proof: Let $\mathcal{S}_i$, $i \in I$, be a countable subbase of $\mathcal{U}_i$ with $X \in \mathcal{S}_i$. The system $\mathcal{S} := \{\bigcap_I S_{ij_i} : S_{ij_i} \in \mathcal{S}_i^\cap,\ i \in I\}$ forms a countable base of $\bigtimes_I \mathcal{U}_i$. The proof is concluded if we show that each set of $\mathcal{S}$ is σ-compact in the product topology. As $\mathcal{U}_i$ is tight, $S_{ij_i} = \bigcup_{l=1}^{\infty} K_{il}$ for each $S_{ij_i} \in \mathcal{S}_i^\cap$, where K_{il} is compact in the topology $\mathcal{U}_1$. Hence, each $S \in \mathfrak{S}$ is the countable union of sets $\bigcap_I K_{il_i}$, which are compact with respect to the product topology, as was shown in (4.11).

5. Perfect Measures

Let $P/\mathcal{A}^*$ be a p-measure. Let R^1 be the real line and $\mathcal{B}^1$ its Borel - algebra (with respect to the usual topology). Let $f: X \to R^1$ be $\mathcal{A}^*, \mathcal{B}^1$ - measurable. Let $\mathcal{A}_f^* = \{D \subset R^1 : f^{-1}D \in \mathcal{A}^*\}$. Then $\mathcal{B}^1 \subset \mathcal{A}_f^*$. Let $P_f/\mathcal{A}_f^*$ be the p-measure induced by f and $P/\mathcal{A}^*$. Following Gnedenko and Kolmogorov (see Gnedenko and Kolmogorov, p. 18 - 19) we define:

Definition (5.1) The measure $P/\mathcal{A}^*$ is perfect, iff $\mathcal{B}^1$ $P_f/\mathcal{A}_f^*$ - approximates $\mathcal{A}_f^*$ for each $\mathcal{A}^*, \mathcal{B}^1$ - measurable f.

According to (2.4) this is the case iff to each $D \in \mathcal{A}_f^*$ there exists $B \in \mathcal{B}^1$ such that $B \subset D$ and $P_f(B) = P_f(D)$.

Remark: Perfectness of $P/\mathcal{A}^*$ implies perfectness of $P/\mathcal{A}_0^*$ for any sub-σ-algebra $\mathcal{A}_0^* \subset \mathcal{A}^*$.

Criterion (5.2) The measure $P/\mathcal{A}^*$ is perfect iff $\mathcal{B}^1$ $P_f/\mathcal{A}_f^*$ - approximates $f(X)$ for each $\mathcal{A}^*,\mathcal{B}^1$ - measurable f.

Proof: As $f(X) \in \mathcal{A}_f^*$, necessity is obvious. To prove sufficiency, let $D \in \mathcal{A}_f^*$. If $D = R^1$, $P_f(f(X)) = P_f(D)$ implies that $\mathcal{B}^1$ also $P_f/\mathcal{A}_f^*$ - approximates D. If $D \neq R^1$, define

$$g(x) := \begin{cases} f(x) & x \in f^{-1}D \\ c & x \in (f^{-1}D)^c \text{ with } c \in D^c. \end{cases}$$

g is $\mathcal{A}^*,\mathcal{B}^1$ - measurable. Therefore there exists $B \in \mathcal{B}^1$ such that $B \subset g(X)$ and $P_f(B) = P_f(g(X))$. As $g(X) = f(f^{-1}D) + \{c\}$, we have $B - \{c\} \subset f(f^{-1}D) \subset D$, and $P_f(B) = P_f(f(f^{-1}D)) + P_f(\{c\}) = P_f(D) + P_f(\{c\})$. Furthermore, $B - \{c\} \in \mathcal{B}^1$. Hence, $\mathcal{B}^1$ $P_f/\mathcal{A}_f^*$ - approximates D.

Theorem (5.3) (i) If $\mathcal{A}^*$ is compact $P/\mathcal{A}^*$ - approximable, then $P/\mathcal{A}^*$ is perfect.
(ii) If $P/\mathcal{A}^*$ is perfect and $\mathcal{A}^*$ is separable, then it is compact $P/\mathcal{A}^*$ - approximable.

Proof:+) (i) Let $f: X \to R^1$ be $\mathcal{A}^*, \mathcal{B}^1$ - measurable. We will show that to each $\varepsilon > 0$ there exists a set X_ε such that $f(X_\varepsilon) \in \mathcal{B}^1$ and $P(X_\varepsilon) > 1 - \varepsilon$.

Let $(J_n)_{n=1,2,\ldots}$ be the system of all rational intervals. As f is $\mathcal{A}^*, \mathcal{B}^1$ - measurable, we have $f^{-1}J_n \in \mathcal{A}^*$ for $n = 1,2,\ldots$. As $\mathcal{A}^*$ is compact $P/\mathcal{A}^*$ - approximable, there exists a compact system $\mathcal{C} \subset \mathcal{A}^*$ which $P/\mathcal{A}^*$ - approximates $\mathcal{A}^*$ (see (3.6)). Therefore to any $\varepsilon > 0$ and each $n = 1,2,\ldots$ there exist sets C_n, $C_n' \in \mathcal{C}$ such that $C_n \subset f^{-1}J_n$, $C_n' \subset f^{-1}J_n^c$, $P(f^{-1}J_n - C_n) < \frac{\varepsilon}{2^{n+1}}$ and $P(f^{-1}J_n^c - C_n') < \frac{\varepsilon}{2^{n+1}}$. Let $X_\varepsilon := \bigcap_1^\infty (C_n \cup C_n')$. Then, $X_\varepsilon \in \mathcal{C}^{\cup\delta}$ and $X_\varepsilon \cap f^{-1}J_n = X_\varepsilon \cap C_n \in \mathcal{C}^{\cup\delta}$. Thus, according to (1.2) - (1.4), $(X_\varepsilon \cap f^{-1}J_n)_{n=1,2,\ldots}$ is a compact system. Furthermore we have

$$P(X_\varepsilon) = 1 - P(\bigcup_1^\infty (C_n^c \cap C_n'^c)) = 1 - P(\bigcup_1^\infty (f^{-1}J_n - C_n) \cup (f^{-1}J_n^c - C_n')) \geq$$
$$\geq 1 - \sum_1^\infty [P(f^{-1}J_n - C_n) + P(f^{-1}J_n^c - C_n')] > 1 - \varepsilon.$$

To conclude the proof it has to be shown that $f(X_\varepsilon) \in \mathcal{B}^1$. We show: $f(X_\varepsilon)$ is closed. Let y be an accumulation point of $f(X_\varepsilon)$ and $(J_{n_k})_{k=1,2,\ldots}$ a decreasing sequence of rational intervals such that $\bigcap_{k=1}^\infty J_{n_k} = \{y\}$. Then $f(X_\varepsilon) \cap J_{n_k} \neq \emptyset$ for $k=1,2,\ldots$.

+) See R y l l - N a r d z e w s k i , Theorem I, p. 126, and Theorem II, p. 127.

Therefore $X_\varepsilon \cap f^{-1}J_{n_k} \neq \emptyset$, $k = 1,2,\ldots$. As $(X_\varepsilon \cap f^{-1}J_{n_k})_{k=1,2,\ldots}$ is again a decreasing sequence, the compactness of $(X_\varepsilon \cap f^{-1}J_n)_{n=1,2,\ldots}$ implies $\bigcap_1^\infty (X_\varepsilon \cap f^{-1}J_{n_k}) = X_\varepsilon \cap f^{-1} \bigcap_1^\infty J_{n_k} = X_\varepsilon \cap f^{-1}\{y\} \neq \emptyset$. Hence, $y \in f(X_\varepsilon)$, q.e.d.

(ii) If $\mathcal{A}^*$ is separable, there exists a function $f: X \to R^1$, such that $\mathcal{A}^* = f^{-1}\mathcal{B}^1$ (see (0.5)). According to (5.1) there exists a set $Y_0 \in \mathcal{B}^1$, $Y_0 \subset f(X)$ and $P_f(Y_0) = 1$. Furthermore, $\mathcal{B}^1$ is compact $P_f/\mathcal{B}^1$ - approximable (see (4.3)). Hence $\mathcal{A}^*$ is compact $P/\mathcal{A}^*$ - approximable according to (3.7).

Criterion (5.4) The measure $P/\mathcal{A}^*$ is perfect iff each separable sub-σ-algebra $\mathcal{A}_0^*$ is compact $P/\mathcal{A}_0^*$ - approximable.

Proof: Necessity. Let $\mathcal{A}_0^* \subset \mathcal{A}^*$ be separable. As $P/\mathcal{A}_0^*$ is perfect, $\mathcal{A}_0^*$ is compact $P/\mathcal{A}_0^*$ - approximable according to (5.3(ii)).

Sufficiency. Let f be $\mathcal{A}^*,\mathcal{B}^1$ - measurable. Then, $f^{-1}\mathcal{B}^1$ is a separable sub-σ-algebra of $\mathcal{A}^*$ which is compact $P/f^{-1}\mathcal{B}^1$-approximable by assumption. From (5.3(i)) we obtain that $P/f^{-1}\mathcal{B}^1$ is perfect. Therefore, there exists a set $Y_0 \in \mathcal{B}^1$, $Y_0 \subset f(X)$ such that $P_f(Y_0) = 1$. Hence, $\mathcal{B}^1$ $P_f/\mathcal{A}_f^*$ - approximates $f(X)$. According to (5.2), this implies perfectness of $P/\mathcal{A}^*$.

6. Existence of Product Measures

In this chapter we will prove two generalizations (due to R y l l - N a r d z e w s k i and M a r c z e w s - k i) of a well known theorem of K o l m o g o r o v on the existence of measure in a product space.

Let I be an arbitrary index set and $\mathfrak{J}_0$ the system of all finite subsets of I. Let $(\mathcal{A}_i^*)_{i\in I}$ be a family of algebraically σ-independent σ-algebras over X. We will use the following notations:

$$\mathcal{A}_{I_0} := \alpha(\bigcup_{I_0} \mathcal{A}_i^*) \qquad \text{for any } I_0 \subset I.$$
$$\mathcal{A}_{I_0}^* := \sigma(\bigcup_{I_0} \mathcal{A}_i^*)$$

For abbreviation, we will write $\mathcal{A}$ instead of $\mathcal{A}_I$ and $\mathcal{A}^*$ instead of $\mathcal{A}_I^*$. Furthermore, we will denote $\bar{\mathcal{A}} := \bigcup_{\mathfrak{J}_0} \mathcal{A}_{I_0}^*$. We remark that $\bar{\mathcal{A}}$ is an algebra (see (0.1)) and that $\mathcal{A} \subset \bar{\mathcal{A}} \subset \mathcal{A}^*$.

Lemma (6.1) Let $P/\mathcal{A}$ [$P/\mathcal{A}^*$] be a p-content [-measure] such that for each $i\in I$ the algebra $\mathcal{B}_i \subset \mathcal{A}_i^*$ is $P/\mathcal{B}_i$ - approximable by the compact system $\mathcal{C}_i \subset \mathcal{A}_i^*$. Then $\alpha(\bigcup_I \mathcal{B}_i)$ [$\sigma(\bigcup_I \mathcal{B}_i)$] is $P/\alpha(\bigcup_I \mathcal{B}_i)$ - [$P/\sigma(\bigcup_I \mathcal{B}_i)$ -] approximable by the compact system $(\bigcup_I \mathcal{C}_i)^{\cup\cap}$ [$(\bigcup_I \mathcal{C}_i)^{\cup\delta}$].

Proof: According to (2.12) we have that $(\bigcup_I \mathcal{C}_i)^{\cap\cup} = (\bigcup_I \mathcal{C}_i)^{\cup\cap}$ $P/\alpha(\bigcup_I \mathcal{B}_i)$ - approximates $\alpha(\bigcup_I \mathcal{B}_i)$. Thus, according to (2.10) we obtain for the case of a p-measure $P/\mathcal{A}^*$ that $(\bigcup_I \mathcal{C}_i)^{\cup\delta}$ $P/\sigma(\bigcup_I \mathcal{B}_i)$ - approximates $\sigma(\bigcup_I \mathcal{B}_i)$. As $\mathcal{A}_i^*$, $i \in I$, are alg. σ-independent, $\mathcal{C}_i$, $i \in I$, are alg. σ-independent too. Hence, $(\bigcup_I \mathcal{C}_i)^{\cup\cap} [(\bigcup_I \mathcal{C}_i)^{\cup\delta}]$ is compact according to (1.6), (1.3) and (1.4).

<u>Theorem (6.2)</u> Let $P/\mathcal{A}$ be a p-content such that $\mathcal{A}_i^*$ is compact $P/\mathcal{A}_i^*$ - approximable for each $i \in I$. Then, $\mathcal{A}$ is compact $P/\mathcal{A}$ - approximable and therefore $P/\mathcal{A}$ σ-additive. Furthermore, $\mathcal{A}^*$ is compact $P/\mathcal{A}^*$ - approximable (where $P/\mathcal{A}^*$ is the (uniquely determined) extension of $P/\mathcal{A}$).

<u>Proof:</u> As $\mathcal{A}_i^*$ is compact $P/\mathcal{A}_i^*$ - approximable, $P/\mathcal{A}_i^*$ is a measure (see (3.4)) and therefore there exists a compact system $\mathcal{C}_i \subset \mathcal{A}_i^*$ $P/\mathcal{A}_i^*$ - approximating $\mathcal{A}_i^*$ according to (3.6). Thus the assumptions of (6.1), version 1, are fulfilled for $\mathcal{B}_i = \mathcal{A}_i^*$, $i \in I$. Hence, $\mathcal{A}$ is compact $P/\mathcal{A}$ - approximable and therefore $P/\mathcal{A}$ is σ-additive. Compact approximation of $\mathcal{A}^*$ follows from (3.5).

<u>Theorem (6.3)</u> Let $P/\bar{\mathcal{A}}$ be a p-content such that

(i) $P/\mathcal{A}_{I_0}^*$ is a measure for each $I_0 \in \mathfrak{J}_0$,

(ii) $\mathcal{A}_i^*$ is compact $P/\mathcal{A}_i^*$ - approximable for each $i \in I$.

Then, $\bar{\mathcal{A}}$ is compact $P/\bar{\mathcal{A}}$ - approximable and therefore $P/\bar{\mathcal{A}}$ is σ-additive.

<u>Proof:</u> Let $\mathcal{C}_i \subset \mathcal{A}_i^*$ be a compact system $P/\mathcal{A}_i^*$ - approximating $\mathcal{A}_i^*$. From (6.1), version 2, applied for I_0 and $\mathcal{A}_{I_0}^*$ instead of I and $\mathcal{A}^*$ and with $\mathcal{B}_i = \mathcal{A}_i^*$ we obtain that $\mathcal{A}_{I_0}^*$ is $P/\mathcal{A}_{I_0}^*$ - approximable by $(\bigcup_{I_0} \mathcal{C}_i)^{\cup\delta}$. Hence, each $\mathcal{A}_{I_0}^*$

and therefore $\bar{\mathcal{A}} = \bigcup_{\mathfrak{J}_0} \mathcal{A}^*_{I_0}$ are compact $P/\bar{\mathcal{A}}$ - approximable by the compact system $(\bigcup_I \mathcal{C}_i)^{\cup\delta}$.

As a consequence of (6.3), $P/\bar{\mathcal{A}} = P^*/\bar{\mathcal{A}}$, where P^* is the extension of $P/\mathcal{A}$. Hence (6.2) implies that $\mathcal{A}^*$ is compact $P/\mathcal{A}^*$ - approximable, where $P/\mathcal{A}^*$ is the extension of $P/\bar{\mathcal{A}}$ ($P/\mathcal{A}^* = P^*/\mathcal{A}^*$).

<u>Lemma (6.4)</u> Let $P/\mathcal{A}$ [$P/\mathcal{A}^*$] be a p-content [-measure] such that $P/\mathcal{A}_i^*$ is a perfect measure for each $i \in I$. Then, each countable sub-algebra $\mathcal{A}_0 \subset \mathcal{A}$ [separable sub-σ-algebra $\mathcal{A}_0^* \subset \mathcal{A}^*$] is compact $P/\mathcal{A}$ - [$P/\mathcal{A}_0^*$ -] approximable.

<u>Proof:</u> Let $\mathcal{A}_0$ [$\mathcal{A}_0^*$] be a countable sub-algebra of $\mathcal{A}$ [separable sub-σ-algebra of $\mathcal{A}^*$]. From (0.3) we have $\mathcal{A}_0 \subset \alpha(\bigcup_I \mathcal{S}_i)$ [$\mathcal{A}_0^* \subset \sigma(\bigcup_I \mathcal{S}_i)$] with countable $\mathcal{S}_i \subset \mathcal{A}_i^*$, $i \in I$. Perfectness of $P/\mathcal{A}_i^*$ together with (5.4) imply that for each $i \in I$ $\sigma(\mathcal{S}_i)$ is compact $P/\sigma(\mathcal{S}_i)$ - approximable. Therefore, according to (3.6), there exists a compact system $\mathcal{C}_i \subset \sigma(\mathcal{S}_i)$ $P/\sigma(\mathcal{S}_i)$ approximating $\sigma(\mathcal{S}_i)$. Thus the assumptions of (6.1) are fulfilled for $\mathcal{B}_i = \sigma(\mathcal{S}_i)$, $i \in I$. Hence $\alpha(\bigcup_I \sigma(\mathcal{S}_i))$ is compact $P/\alpha(\bigcup_I \sigma(\mathcal{S}_i))$ - approximable [$\sigma(\bigcup_I \sigma(\mathcal{S}_i))$ is compact $P/\sigma(\bigcup_I \sigma(\mathcal{S}_i))$ - approximable]. As $\mathcal{A}_0 \subset \alpha(\bigcup_I \sigma(\mathcal{S}_i)) \subset \mathcal{A}$, the statement about the compact approximation of $\mathcal{A}_0$ is proved. - Because $\sigma(\bigcup_I \sigma(\mathcal{S}_i))$ is compact $P/\sigma(\bigcup_I \sigma(\mathcal{S}_i))$ - approximable, $P/\sigma(\bigcup_I \sigma(\mathcal{S}_i))$ is perfect (see (5.3(i)). Hence $\mathcal{A}_0^* \subset \sigma(\bigcup_I \sigma(\mathcal{S}_i))$ is compact $P/\mathcal{A}_0^*$ - approximable according to (5.4).

Theorem (6.5) Let $P/\mathcal{A}$ be a p-content such that $P/\mathcal{A}_i^*$ is a perfect measure for each $i \in I$. Then $P/\mathcal{A}$ is σ-additive and the (uniquely determined) extension $P/\mathcal{A}^*$ is perfect.

Proof: According to (6.4), version 1, each countable sub-algebra $\mathcal{A}_0 \subset \mathcal{A}$ is compact $P/\mathcal{A}$ - approximable. This, however implies σ-additive of $P/\mathcal{A}$. Therefore, there exists a uniquely determined extension $P/\mathcal{A}^*$. According to (6.4), version 2, each separable sub-σ-algebra $\mathcal{A}_0^*$ of $\mathcal{A}^*$ is compact $P/\mathcal{A}_o^*$ - approximable. According to (5.4) this implies perfectness of $P/\mathcal{A}^*$.

Theorem (6.6) Let $P/\bar{\mathcal{A}}$ be a probability content such that

(i) $P/\mathcal{A}_{I_0}^*$ is a measure for all $I_o \in \mathfrak{I}_o$,

(ii) $P/\mathcal{A}_i^*$ is a perfect measure for all $i \in I$.

Then, $P/\bar{\mathcal{A}}$ is σ-additive.

Proof: According to (0.4) for each countable sub-algebra $\mathcal{A}_0 \subset \bar{\mathcal{A}}$ there exist countable systems $\mathcal{S}_i \subset \mathcal{A}_i^*$ such that $\mathcal{A}_o \subset \bigcup_{\mathfrak{I}_o} \sigma(\bigcup_{I_o} \mathcal{S}_i)$. Perfectness of $P/\mathcal{A}_i^*$ together with (5.4) and (3.6) imply that there exists a compact system $\mathcal{C}_i \subset \sigma(\mathcal{S}_i)$ $P/\sigma(\mathcal{S}_i)$ - approximating $\sigma(\mathcal{S}_i)$ for each $i \in I$. Applying (6.1), version 2, for I_0 and $\mathcal{A}_{I_0}^*$ instead of I and $\mathcal{A}^*$ and with $\mathcal{B}_i = \sigma(\mathcal{S}_i)$, (i) yield that $\sigma(\bigcup_{I_0} \mathcal{S}_i)$ is $P/\mathcal{A}_{I_0}^*$ - approximable by the compact system $(\bigcup_{I_0} \mathcal{C}_i)^{\cup\delta}$. Hence, $\sigma(\bigcup_{I_0} \mathcal{S}_i)$ is $P/\bar{\mathcal{A}}$ - approximable by the compact system $(\bigcup_{I} \mathcal{C}_i)^{\cup\delta}$, for each $I_o \in \mathfrak{I}_o$. Therefore, $\mathcal{A}_o \subset \bigcup_{\mathfrak{I}_o} \sigma(\bigcup_{I_o} \mathcal{S}_i)$ is $P/\bar{\mathcal{A}}$ - approximable by the compact system $(\bigcup_{I} \mathcal{C}_i)^{\cup\delta}$. By (3.4) this implies that P is σ-additive on $\mathcal{A}_0$. As $\mathcal{A}_0$ was an arbitrary countable sub-algebra of $\tilde{\mathcal{A}}$, P is σ-additive on $\tilde{\mathcal{A}}$.

As a consequence of (6.6), $P/\tilde{\mathcal{A}} = P^*/\tilde{\mathcal{A}}$, where P^* is the extension of $P/\mathcal{A}$. Hence the consequence of (6.5) is also valid in this case, i.e. the extension $P/\mathcal{A}^*$ of $P/\tilde{\mathcal{A}}$ is perfect ($P/\mathcal{A}^* \equiv P^*/\mathcal{A}^*$).

Now consider the case that X is a cartesian product: $X = \underset{I}{\times} X_i$. Assume that for each $i \in I$ we have a σ-algebra $\mathcal{B}_i^*$ on the space X_i. Let $Z_{\{i\}}(B) := \{x \in X: x_i \in B\}$ for each $B \in \mathcal{B}_i^*$ and each $i \in I$. Then the σ-algebras $Z_{\{i\}}(\mathcal{B}_i^*) := \{Z_{\{i\}}(B): B \in \mathcal{B}_i^*\}$ are alg. σ-independent. Let $\mathcal{A} := \alpha(\underset{I}{\cup} Z_{\{i\}}(\mathcal{B}_i^*))$ and $\mathcal{A}^* := \sigma(\mathcal{A})$. The results of this section applied to this case lead to:

<u>Corollary (6.7)</u> Let X be a cartesian product, $X = \underset{I}{\times} X_i$, and let $P/\mathcal{A}$ be a p-content. Assume that for each $i \in I$ $\mathcal{B}_i^*$ is compact $P_i/\mathcal{B}_i^*$ - approximable where $P_i(B) := P(Z_{\{i\}}(B))$ for each $B \in \mathcal{B}_i^*$ and each $i \in I$. Then $\mathcal{A}$ is compact $P/\mathcal{A}$ - approximable. (Hence $P/\mathcal{A}$ is σ-additive.)

<u>Proof:</u> Because $P/Z_{\{i\}}(\mathcal{B}_i^*)$ is a p-content, $P_i/\mathcal{B}_i^*$ is a p-content too, for each $i \in I$. As $\mathcal{B}_i^*$ is compact $P_i/\mathcal{B}_i^*$ - approximable, say by the compact system $\mathcal{C}_i$, $Z_{\{i\}}(\mathcal{B}_i^*)$ is $P/Z_{\{i\}}(\mathcal{B}_i^*)$ - approximable by the compact system $Z_{\{i\}}(\mathcal{C}_i)$. Therefore, the corollary follows from (6.2).

As an immediate consequence of (4.3) and (6.7) we obtain:

Corollary (6.8) Let $X = \bigtimes_I X_i$ and assume that over each space X_i a tight topology is defined. Let $\mathcal{B}_i^*$ be the Borel - algebra on X_i and let $P/\mathcal{A}$ be a p-content such that $P/Z_{\{i\}}(\mathcal{B}_i^*)$ is a measure for each $i \in I$. Then $\mathcal{A}$ is compact $P/\mathcal{A}$ - approximable. (Hence $P/\mathcal{A}$ is σ-additive.)

Similar corollaries can be obtained for Theorems (6.3), (6.5) and (6.6).

That σ-additivity of $P/\mathcal{A}$ (or $P/\bar{\mathcal{A}}$) does not follow from σ-additivity of $P/\mathcal{A}_i^*$ for all $i \in I$ (or σ-additivity of $P/\mathcal{A}_{I_0}^*$ for all $I_0 \in \mathfrak{I}_0$) without further assumptions is shown by the following example (essentially due to H a l m o s , p. 214).

Example: Let $Y_0 : = [0,1)$, let $\mathcal{B}_0^*$ be the Borel - algebra over Y_0 (with respect to the usual topology of the real line). Let $Y : = \bigtimes_1 Y_i$ with $Y_i = Y_0$ for $i = 1,2,\dots$. Let

$$\mathcal{B} : = \bigcup_1^\infty \mathcal{B}_n', \quad \mathcal{B}_n' : = \alpha\Big(\bigcup_{i=1}^n Z_{\{i\}}(\mathcal{B}_0^*)\Big)$$
$$\bar{\mathcal{B}} : = \bigcup_1^\infty \mathcal{B}_n'^*, \quad \mathcal{B}_n'^* : = \sigma\Big(\bigcup_{i=1}^n Z_{\{i\}}(\mathcal{B}_0^*)\Big)$$

Let $(X_i)_{i=1,2,\dots}$ be a decreasing sequence of sets with $X_i \subset Y_0$, $\lambda^*(X_i) = 1$ for $i = 1,2,\dots$ and $\bigcap_1^\infty X_i = \emptyset$, where $\lambda/\mathcal{B}_0^*$ is the Lebesgue-measure, λ^* the corresponding outer measure. (The existence of such a sequence can be proved similarly to H a l m o s , pp 68 ff. The detailed proof is too lengthy to be included here.) Let $X : = \bigtimes_1^\infty X_i$ and define:

$$\mathcal{A} : = X \cap \mathcal{B}, \quad \mathcal{A}_n' : = X \cap \mathcal{B}_n'$$
$$\bar{\mathcal{A}} : = X \cap \bar{\mathcal{B}}, \quad \mathcal{A}_n'^* : = X \cap \mathcal{B}_n'^* \quad \text{and } \mathcal{A}_i^* = X \cap Z_{\{i\}}(\mathcal{B}_0^*)$$

We have (H a l m o s , Theorem E, p.25):

$$\mathcal{A} = \bigcup_{1}^{\infty} \mathcal{A}'_n, \quad \mathcal{A}'_n = \alpha(\bigcup_{1}^{n} \mathcal{A}^*_i),$$
$$\bar{\mathcal{A}} = \bigcup_{1}^{\infty} \mathcal{A}'^*_n, \quad \mathcal{A}'^*_n = \sigma(\bigcup_{1}^{n} \mathcal{A}^*_i).$$

Now we define a map $T: Y_0 \to Y$ by

$$T(y) := (y,y,\dots) \quad \text{for each } y \in Y_0.$$

We remark that T is $\mathcal{B}^*_0,\bar{\mathcal{B}}$- measurable : As for each $B \in \mathcal{B}^*_0$, $T^{-1}Z_{\{i\}}(B) = B$, we have $T^{-1}\bar{\mathcal{B}} \subset \mathcal{B}^*_0$. Therefore, we can define a p-content $\lambda_T/\bar{\mathcal{B}}$ by

$$\lambda_T(B) := \lambda(T^{-1}B) \quad \text{for each } B \in \bar{\mathcal{B}}.$$

We will show, (i) that $\lambda_T(B_1) = \lambda_T(B_2)$ if $X \cap B_1 = X \cap B_2$. Thus we can define (ii) a p-content $P/\bar{\mathcal{A}}$ by

$$P(A) := \lambda_T(B) \quad \text{if } A = B \cap X.$$

We will show,(iii) that P is σ-additive on $\mathcal{A}'^*_n$ for $n = 1,2,\dots$, however (iv) <u>not</u> σ-additive on $\mathcal{A}$ (and therefore not σ-additive on $\bar{\mathcal{A}} \supset \mathcal{A}$).

In this example all assumptions of Theorems (6.2) and (6.3) except compact approximation of $\mathcal{A}^*_i$ are fulfilled. Furthermore, all assumptions of Theorems (6.5) and (6.6) except perfectness of $P/\mathcal{A}^*_i$ are fulfilled.

<u>Proofs:</u> (i) Assume that $X \cap B_1 = X \cap B_2$ with $B_1, B_2 \in \bar{\mathcal{B}}$. Hence there exists n such that $B_1, B_2 \in \mathcal{B}'^*_n$, whence there exist sets $B_1^{(n)}, B_2^{(n)} \subset Y_1 \times \dots \times Y_n$ such that $B_i = Z_{\{1,\dots,n\}}(B_i^{(n)})$, $i = 1,2$, where the cylinder $Z_{\{1,\dots,n\}}$ is to be taken over the components $Y_{n+1}, Y_{n+2}, \dots$. Therefore $X \cap B_1 = X \cap B_2$ implies $X_1 \times \dots \times X_n \cap B_1^{(n)} = X_1 \times \dots \times X_n \cap B_2^{(n)}$ and thus

$$Z_{\{1,\dots,n\}}(X_1 \times \dots \times X_n \cap B_1^{(n)}) = Z_{\{1,\dots,n\}}(X_1 \times \dots \times X_n \cap B_2^{(n)}).$$

Applying T^{-1} to both expressions leads to

$$X_n \cap T^{-1}B_1 = X_n \cap T^{-1}B_2,$$

as $T^{-1}Z_{\{1,\dots,n\}}(X_1\times\dots\times X_n) = X_n$. From $(T^{-1}B_1 \Delta T^{-1}B_2) \subset X_n^c$ and $\lambda^*(X_n) = 1$ we obtain $\lambda(T^{-1}B_1) = \lambda(T^{-1}B_2)$.

(ii) $P(X) = \lambda_T(Y) = \lambda(Y_0) = 1$. Furthermore, $P/\bar{\mathcal{A}} \geq 0$. Additivity of $P/\bar{\mathcal{A}}$ follows from additivity of $P/\mathcal{A}_n^{\prime *}$ that is shown in (iii).

(iii) $P/\mathcal{A}_n^{\prime *}$ is σ-additive. Let $(A_j)_{j=1,2,\dots}$ be such that $A_j \in \mathcal{A}_n^{\prime *}$ for $j = 1,2,\dots$ and $A_{j'} \cap A_{j''} = \emptyset$ for $j' \neq j''$. Let $A_j = X \cap B_j$ with $B_j \in \mathcal{B}_n^{\prime *}$ for $j = 1,2,\dots$. Without restriction of generality we can assume $B_{j'} \cap B_{j''} = \emptyset$ for $j' \neq j''$ (otherwise, we take $\hat{B}_j = B_j - (\bigcup_1^{j-1} B_i)$ instead of B_j, as $X \cap \hat{B}_j = X \cap B_j$ if $(X \cap B_{j'}) \cap (X \cap B_{j''}) = \emptyset$ for $j' \neq j''$). Since $B_j \in \mathcal{B}_n^{\prime *}$, we have $\sum_1^\infty B_j \in \mathcal{B}_n^{\prime *}$. Therefore, $\lambda_T(\sum_1^\infty B_j)$ is defined and we have $\lambda_T(\sum_1^\infty B_j) = \sum_1^\infty \lambda_T(B_j)$ which implies σ-additivity of $P/\mathcal{A}_n^{\prime *}$.

(iv) Finally we show that the zero limit theorem does not hold for $P/\mathcal{A}$. Let $D_{nk} := [\frac{k-1}{2^n}, \frac{k}{2^n})$ for $k = 1,\dots,2^n$, $n = 1,2,\dots$. Define $D_n := Z_{\{1,\dots,n\}}(\sum_{k=1}^{2^n} D_{nk}\times\dots\times D_{nk})$ where the cylinder $Z_{\{1,\dots,n\}}$ is taken over the components $Y_{n+1}, Y_{n+2}, \dots$. We have $D_{n+1} \subset D_n$ and $D_n \in \mathcal{B}_n'$ for $n = 1,2,\dots$. Therefore, $A_n := X \cap D_n$ is a non-increasing sequence such that $A_n \in \mathcal{A}_n'$. Furthermore, $\bigcap_1^\infty A_n = X \cap \bigcap_1^\infty D_n = \emptyset$, because

$\bigcap_1^\infty D_n = \{(y,y,\ldots): y \in Y_0\}$ and $\bigcap_1^\infty X_i = \emptyset$, i.e. there exists no y such that $y \in X_i$ for all $i = 1,2,\ldots$. On the other hand $T^{-1}D_n = Y_0$ and therefore $P(A_n) = 1$ for all $n = 1,2,\ldots$.

We remark that V. Baumann has constructed an example of this kind with identical component spaces X_i.

7. Existence of Regular Conditional Probability Measures

In this chapter $P/\mathcal{A}^*$ will always denote a p-measure, $\mathcal{B}^* \subset \mathcal{A}^*$ an arbitrary sub-σ-algebra, and $\mathcal{B}^1$ the Borel - algebra of the real line (with respect to the usual topology).

Definition (7.1) Let $A \in \mathcal{A}^*$. A conditional probability of A given $\mathcal{B}^*$, denoted by $P(A,x/\mathcal{B}^*)$, is a conditional expectation of $\chi_A(x)$ given $\mathcal{B}^*$, i.e.

(i) $P(A,\cdot/\mathcal{B}^*)$ is a $\mathcal{B}^*,\mathcal{B}^1$ - measurable function

(ii) $\int_B P(A,x/\mathcal{B}^*)dP(x) = P(A \cap B)$ for each $B \in \mathcal{B}^*$,

We remark, that conditional probabilities exist for each $A \in \mathcal{A}^*$. For given A, $P(A,\cdot/\mathcal{B}^*)$ is uniquely determined up to $P/\mathcal{B}^*$ - null sets.

It is necessary to consider $P(A,x/\mathcal{B}^*)$ not only as a function of x but also as a function of $A \in \mathcal{A}^*$: $P(\cdot,\cdot/\mathcal{B}^*)/\mathcal{A}^* \times X$.

From the properties of conditional expectations (see e.g. L o è v e , pp 347-348) we obtain the following properties of $P(\cdot,\cdot/\mathcal{B}^*)$:

(7.2) $P(X,x/\mathcal{B}^*) = 1$ and $P(\emptyset,x/\mathcal{B}^*) = 0$ $P/\mathcal{B}^*$ - a.e.

(7.3) $0 \leq P(A,x/\mathcal{B}^*) \leq 1$ $P/\mathcal{B}^*$ - a.e. for each $A \in \mathcal{A}^*$.

(7.4) $P(\sum_1^\infty A_i,x/\mathcal{B}^*) = \sum_1^\infty P(A_i,x/\mathcal{B}^*)$ $P/\mathcal{B}^*$ - a.e. for each sequence $(A_i)_{i=1,2,\ldots}$ of disjoint sets.

(7.5) $P(A_1,x/\mathcal{B}^*) \leq P(A_2,x/\mathcal{B}^*)$ $P/\mathcal{B}^*$ - a.e. for $A_1 \subset A_2$.

Properties (7.2) - (7.4) suggest that $P(.,./\mathcal{B}^*)$ might be chosen such that $P(.,x/\mathcal{B}^*)$ for fixed x considered as a function of A will be a p-measure. This is however not the case in general because the exceptional sets in (7.2) - (7.4) might depend on the sets $A \in \mathcal{A}^*$ respectively the sequences $(A_i)_{i=1,2,...} \subset \mathcal{A}^*$.

<u>Definition (7.5)</u> Let $\mathcal{A}_0^* \subset \mathcal{A}^*$ be a sub-σ-algebra. $P(.,./\mathcal{B}^*)$ is <u>regular</u> on $\mathcal{A}_0^*$, iff $P(.,x/\mathcal{B}^*)/\mathcal{A}_0^*$ is a p-measure for each $x \in X$.

The following example due to D i e u d o n n é shows that even in the case of a separable σ-algebra regular conditional probabilities do not necessarily exist.

<u>Example:</u> Let $X = [0,1]$, $\mathcal{B}^*$ the Borel - algebra on X and $\lambda/\mathcal{B}^*$ the Lebesgue measure. Let $M \subset X$ have outer Lebesgue measure 1 and inner Lebesgue measure 0. (For the existence of a set with these properties see e.g. H a l m o s , theorem E, p.70.).

Define:

$$\mathcal{A}^* := \{(B' \cap M) \cup (B'' \cap M^c) : B', B'' \in \mathcal{B}^*\}$$

and

$$P((B' \cap M) \cup (B'' \cap M^c)) := \lambda(B').$$

Then no regular conditional probability on $\mathcal{A}^*$ given $\mathcal{B}^*$ exists.

<u>Proof:</u> Obviously, $\mathcal{B}^* \subset \mathcal{A}^*$ (for $B = (B \cap M) \cup (B \cap M^c)$ for each $B \in \mathcal{B}^*$) and $\mathcal{A}^*$ is a σ-algebra.

We show that the definition of $P/\mathcal{A}^*$ is unique. Let $(B_1' \cap M) \cup (B_1'' \cap M^c) = (B_2' \cap M) \cup (B_2'' \cap M^c)$. This implies $B_1' \cap M = B_2' \cap M$, whence $B_1' \Delta B_2' \subset M^c$. As M has outer Lebesgue measure 1, we therefore obtain $\lambda(B_1') = \lambda(B_2')$. Thus $P((B_1' \cap M) \cup (B_1'' \cap M^c)) = P((B_2' \cap M) \cup (B_2'' \cap M^c))$ by definition.

Of course, $P(A) \geq 0$ for each $A \in \mathcal{A}^*$ and $P(X) = P((X \cap M) \cup (X \cap M^c)) = \lambda(X) = 1$. Let $A_i \in \mathcal{A}^*$, $i = 1,2,\ldots$ with $A_i = (B_i' \cap M) \cup (B_i'' \cap M^c)$ and $A_i \cap A_j = \emptyset$ for $i \neq j$. Then $(B_i' \cap M) \cap (B_j' \cap M) = \emptyset$ for $i \neq j$, that means $B_i' \cap B_j' \subset M^c$ for $i \neq j$. Therefore we have:

$\lambda(B_i' \cap B_j') = 0$ for $i + j$. Thus $P(\overset{\infty}{\underset{1}{\cup}} A_i) = \lambda(\overset{\infty}{\underset{1}{\cup}} B_i') = \lim\limits_{n\to\infty} \lambda(\overset{n}{\underset{1}{\cup}} B_i') = = \lim\limits_{n\to\infty} \sum\limits_1^n \lambda(B_i') = \sum\limits_1^\infty \lambda(B_i')$. Hence, $P/\mathcal{A}^*$ is a p - measure.

Now assume that there exists a regular conditional probability on $\mathcal{A}^*$ given $\mathcal{B}^*$, say $P(.,./\mathcal{B}^*)$. Then there exists a set $N \in \mathcal{B}^*$ with $P(N)(= \lambda(N)) = 0$ such that for each $x \in N^c$

$$(+) \qquad P(B,x/\mathcal{B}^*) = \chi_B(x) \quad \text{for all } B \in \mathcal{B}^*.$$

The proof of this statement rests on the separability of $\mathcal{B}^*$. Let $\mathcal{D} = \{D_1, D_2, \ldots\}$ be a countable algebra such that $\mathcal{B}^* = \sigma(\mathcal{D})$. As $P(A,x/\mathcal{B}^*)$ is - for each $A \in \mathcal{A}^*$ - a conditional expectation of $\chi_A(x)$ with respect to $\mathcal{B}^*$, $D_i \in \mathcal{B}^*$ implies that for each $i = 1,2,\ldots$ there exists a set $N_i \in \mathcal{B}^*$ with $P(N_i) = 0$ such that

$$P(D_i,x/\mathcal{B}^*) = \chi_{D_i}(x) \quad \text{for all } x \in N_i^c.$$

Let $N := \overset{\infty}{\underset{1}{\cup}} N_i$. Then, $N \in \mathcal{B}^*$, $P(N) = 0$ and for each $x \in N^c$ we have

$$P(D_i,x/\mathcal{B}^*) = \chi_{D_i}(x) \quad \text{for all } i = 1,2,\ldots$$

For $x \in N^c$, $P(.,x/\mathcal{B}^*)$ and $\chi_{.}(x)$ are p-measures coinciding on $\mathcal{S}$. According to the extension theorem they coincide on $\mathcal{B}^*$, which proves our statement.

As $\{x\} \in \mathcal{B}^*$, (+) implies

$$P(\{x\},x/\mathcal{B}^*) = \chi_{\{x\}}(x) = 1 \quad \text{for all } x \in N^c.$$

Furthermore, as $P(M) = 1$, there exists a set $N_0 \in \mathcal{B}^*$, with $P(N_0) = 1$ such that

$$P(M,x/\mathcal{B}^*) = 1 \quad \text{for all } x \in N_0^c.$$

For each p-measure P, $P(A) = 1$ and $P(B) = 1$ together imply $P(A \cap B) = 1$. Thus,

$$P(\{x\} \cap M,x/\mathcal{B}^*) = 1 \quad \text{for all } x \in (N \cup N_0)^c.$$

$\{x\} \cap M$ is either $\{x\}$ or $\emptyset$. Because (+) implies $P(\emptyset,x/\mathcal{B}^*) = 0$ for all $x \in N^c$, we have $\{x\} \cap M = \{x\}$ for all $x \in (N \cup N_0)^c$, i.e. $(N \cup N_0)^c \subset M$. As $(N \cup N_0)^c \in \mathcal{B}^*$, $P((N \cup N_0)^c) = 1$ implies $\lambda((N \cup N_0)^c) = 1$ which contradicts the assumption that M has inner Lebesgue measure 0.

Following J i ř i n a (Theorem III, p.82) we will now show that for each separable sub-σ-algebra $\mathcal{A}_0^* \subset \mathcal{A}^*$ which is compact $P/\mathcal{A}^*$ - approximable and each sub-σ-algebra $\mathcal{B}^* \subset \mathcal{A}^*$ a regular conditional probability on $\mathcal{A}_0^*$ given $\mathcal{B}^*$ exists.

Lemma (7.7) Let $\mathcal{S} \subset \mathcal{A}^*$ be a countable system $P/\mathcal{A}^*$ - approximating the set $A \in \mathcal{A}^*$. Then for each conditional probability $P(.,./\mathcal{B}^*)$ we have

$$P(A,x/\mathcal{B}^*) = \sup\{P(S,x/\mathcal{B}^*): A \supset S \in \mathcal{S}\} \quad P/\mathcal{B}^* \text{ - a.e.}$$

<u>Proof:</u> By assumption

(+) $P(A) = \sup\{P(S): A \supset S \in \mathcal{S}\}$.

According to (7.5) we have for each $S \subset A$

$$P(S,x/\mathcal{B}^*) \leq P(A,x/\mathcal{B}^*) \quad P/\mathcal{B}^* \text{ - a.e.}$$

As $\mathcal{S}$ is countable, this implies

(++) $P(S,x/\mathcal{B}^*) \leq s(x) \leq P(A,x/\mathcal{B}^*) \quad P/\mathcal{B}^*$ - a.e.

where $s(x) := \sup\{P(S,x/\mathcal{B}^*): A \supset S \in \mathcal{S}\}$. Obviously s is $\mathcal{B}^*,\mathcal{B}^1$ - measurable. Integration of (++) with respect to $P/\mathcal{B}^*$ yields

$$P(S) \leq \int s dP \leq P(A).$$

As this relation holds for all $S \in \mathcal{S}$ with $S \subset A$, (+) implies

$$\int s dP = P(A).$$

Together with (++), we obtain

$$s(x) = P(A,x/\mathcal{B}^*) \quad P/\mathcal{B}^* \text{ - a.e.}$$

<u>Theorem (7.8)</u> If $\mathcal{A}_0^* \subset \mathcal{A}^*$ is a separable sub-σ-algebra which is compact $P/\mathcal{A}^*$ - approximable, then there exists a regular conditional probability on $\mathcal{A}_0^*$ given $\mathcal{B}^*$ for any sub-σ-algebra $\mathcal{B}^* \subset \mathcal{A}^*$.

<u>Proof:</u> As $\mathcal{A}_0^*$ is separable, there exists a countable subsystem $\mathcal{A}_1 = \{A_1, A_2, \ldots\}$ generating $\mathcal{A}_0^*$. Because the smallest algebra containing a countable system is countable itself, $\mathcal{A}_1$ can be assumed to be algebra without restriction of generality. By assumption there exists a compact system $\mathcal{C}$ $P/\mathcal{A}^*$ - approximating $\mathcal{A}_0^*$. Therefore we can find for each $A_n \in \mathcal{A}_1$ sequences $(A_{n,k})_{k=1,2,\ldots} \subset \mathcal{A}^*$ and $(C_{n,k})_{k=1,2,\ldots} \subset \mathcal{C}$

such that $A_{n,k} \subset C_{n,k} \subset A_n$ and $P(A_n - A_{n,k}) < \frac{1}{k}$ for each $k = 1,2,\ldots$ (see Remark (i), p.5). Let $\mathcal{A}_2 := \alpha(\mathcal{A}_1 \cup (A_{n,k})_{n,k=1,2,\ldots})$. Of course, $\mathcal{A}_2 \subset \mathcal{A}^*$ is countable.

Let $\mathcal{B}^*$ be an arbitrary sub-σ-algebra of $\mathcal{A}^*$ and choose a conditional probability given $\mathcal{B}^*$, say $P_0(.,./\mathcal{B}^*)$. Since $\mathcal{A}_2$ is countable, according to (7.2) - (7.4) and Lemma (7.7) we can find $P/\mathcal{B}^*$ - null sets N_i, $i = 1,2,3,4$ such that

(i) for all $x \notin N_1$

$$0 \leq P_0(A,x/\mathcal{B}^*) \leq 1 \quad \text{for all } A \in \mathcal{A}_2$$

(ii) for all $x \notin N_2$

$$P_0(X,x/\mathcal{B}^*) = 1$$

(iii) for all $x \notin N_3$

$$P_0(A' + A'',x/\mathcal{B}^*) = P_0(A',x/\mathcal{B}^*) + P_0(A'',x/\mathcal{B}^*)$$

for all disjoint $A',A'' \in \mathcal{A}_2$

(iv) for all $x \notin N_4$

$$P_0(A_n,x/\mathcal{B}^*) = \sup_k P_0(A_{n,k},x/\mathcal{B}^*) \quad \text{for all } A_n \in \mathcal{A}_1.$$

We define for each $A \in \mathcal{A}^*$:

$$P_1(A,x) := \begin{cases} P_0(A,x/\mathcal{B}^*) & \text{for } x \notin \bigcup_1^4 N_i \\ P(A) & \text{for } x \in \bigcup_1^4 N_i \end{cases}$$

Thus, $P_1(.,.)$ is a conditional probability given $\mathcal{B}^*$ and for each $x \in X$ the function $P_1(.,x)$ is a p-content on $\mathcal{A}_2$. As for each $x \in X$ $\mathcal{A}_1$ is $P_1(.,x)/\mathcal{A}_2$ - approximable by the compact system $\mathcal{C}$, $P_1(.,x)/\mathcal{A}_1$ is σ-additive (for each $x \in X$) according to (3.4). Therefore, for each $x \in X$ there exists a p-measure, say $P_1^*(.,x)$, on $\mathcal{A}_o^* = \sigma(\mathcal{A}_1)$, which is the extension of $P_1(.,x)/\mathcal{A}_1$.

It remains to be shown that $P_1^*(.,.)$ is a conditional probability on $\mathcal{A}_o^*$ given $\mathcal{B}^*$. Let $\mathcal{A}_*$ be the system of all sets $A \in \mathcal{A}_o^*$ such that $P_1^*(A,.)$ is a conditional expectation of $\chi_A(x)$ with respect to $\mathcal{B}^*$ (given $P/\mathcal{A}^*$), i.e. the system of all sets A such that

(v) $P_1^*(A,.)$ is $\mathcal{B}^*,\mathcal{B}^1$ - measurable

(vi) $\int_B P_1^*(A,x)dP(x) = P(A \cap B)$ for all $B \in \mathcal{B}^*$.

By definition $\mathcal{A}_1 \subset \mathcal{A}_*$. We will show that $\mathcal{A}_*$ is a monotone system. Then, $\mathcal{A}_* = \mathcal{A}_o^*$ (see Halmos, Theorem B, p.27),

Let $(A_n)_{n=1,2,\ldots}$ be a non-decreasing sequence of sets $A_n \in \mathcal{A}^*$. Then, $P_1^*(\bigcup_1^\infty A_n,.) = \lim_{n\to\infty} P_1^*(A_n,.)$ is $\mathcal{B}^*,\mathcal{B}^1$ - measurable, because each $P_1^*(A_n,.)$ is $\mathcal{B}^*,\mathcal{B}^1$ - measurable. Furthermore, using the monotone convergence theorem, we obtain:

$$\int_B P_1^*(\bigcup_1^\infty A_n,x)dP(x) = \int_B \lim_{n\to\infty} P_1^*(A_n,x)dP(x) = \lim_{n\to\infty} \int_B P_1^*(A_n,x)dP(x) =$$
$$= \lim_{n\to\infty} P(A_n \cap B) = P((\bigcup_1^\infty A_n) \cap B)$$

for each $B \in \mathcal{B}^*$. In the same manner you show that for any non-increasing sequence $(A_n)_{n=1,2,\ldots} \subset \mathcal{A}_*$ the set $\bigcap_{n=1}^\infty A_n$ belongs to $\mathcal{A}_*$, which concludes the proof.

Using (4.3) we obtain

<u>Corollary (7.9)</u> Let $\mathcal{A}_o^*$ be the Borel - algebra of a tight topological Hausdorff space with countable base.

Let $P/\mathcal{A}^*$ be a p-measure on a σ-algebra $\mathcal{A}^* \supset \mathcal{A}_0^*$. Then for any sub-$\sigma$-algebra $\mathcal{B}^*$ of $\mathcal{A}^*$ a regular conditional probability on $\mathcal{A}_0^*$ given $\mathcal{B}^*$ exists.

According to (3.7) we have the following generalization of a theorem of Doob (see Doob, Theorem (9.5), p.31):

<u>Corollary (7.10)</u> Let $\mathcal{D}^*$ be a separable σ-algebra on the space Y and let $T: X \to Y$ be $\mathcal{A}^*,\mathcal{D}^*$ - measurable. Assume that $\mathcal{D}^*$ is compact $P_T/\mathcal{D}^*$ - approximable, where $P_T/\mathcal{D}^*$ is the measure induced by T and $P/\mathcal{A}^*$. Assume further, that T(X) is $P_T/\mathcal{D}^*$ - approximable by $\mathcal{D}^*$. Then for each sub-σ-algebra $\mathcal{B}^*$ of $\mathcal{A}^*$ there exists a regular conditional probability on $T^{-1}\mathcal{D}^*$ given $\mathcal{B}^*$.

<u>Remark:</u> If $\mathcal{D}^*$ is the Borel - algebra of a topological Hausdorff space with countable base, then $\mathcal{D}^*$ is separable and by (4.3) compact $P_T/\mathcal{D}^*$ - approximable for any probability measure $P/\mathcal{A}^*$ and any $\mathcal{A}^*,\mathcal{D}^*$ - measurable map $T: X \to Y$.

As a special case of Corollary (7.10) we have (see also (5.4)):

<u>Corollary (7.11)</u> If the probability measure $P/\mathcal{A}^*$ is perfect, then there exists a regular conditional probability on any separable sub-σ-algebra of $\mathcal{A}^*$ with respect to any arbitrary sub-σ-algebra of $\mathcal{A}^*$.

0. Appendix : Lemmata on the Generation of Algebras

Lemma (0.1) If $(J,<)$ is a (upwardly) directed index set and if for each $j \in J$ an algebra $\mathcal{A}_j$ is given such that $j_1 < j_2$ implies $\mathcal{A}_{j_1} \subset \mathcal{A}_{j_2}$, then $\bigcup_J \mathcal{A}_j$ is an algebra.

Proof: We have to show that $\bigcup_J \mathcal{A}_j$ is closed under complementation and intersection. Let $A \in \bigcup_J \mathcal{A}_j$. Then, $A \in \mathcal{A}_{j_0}$, whence $A^c \in \mathcal{A}_{j_0} \subset \bigcup_J \mathcal{A}_j$. Let $A_1, A_2 \in \bigcup_J \mathcal{A}_j$. Then, $A_1 \in \mathcal{A}_{j_1}$, $A_2 \in \mathcal{A}_{j_2}$. As J is directed, there exists j_0 such that $j_i < j_0$, $i = 1,2$. Hence, $A_i \in \mathcal{A}_{j_i} \subset \mathcal{A}_{j_0}$ for $i = 1,2$. We have $A_1 \cap A_2 \in \mathcal{A}_{j_0} \subset \bigcup_J \mathcal{A}_j$.

Let I be an arbitrary index set and $\mathfrak{J}_0$ the system of all finite subsets of I.

Corollary (0.2) If $(\mathcal{A}_i)_{i \in I}$ is a family of algebras on X, then $\alpha(\bigcup_I \mathcal{A}_i) = \bigcup_{\mathfrak{J}_0} (\bigcup_{I_0} \mathcal{A}_i)^{\cap \cup}$.

Proof: From $(\bigcup_{I_0} \mathcal{A}_i)^{\cap \cup} \subset \alpha(\bigcup_I \mathcal{A}_i)$ for each subset $I_0 \subset I$ we immediately obtain $\bigcup_{\mathfrak{J}_0} (\bigcup_{I_0} \mathcal{A}_i)^{\cap \cup} \subset \alpha(\bigcup_I \mathcal{A}_i)$.

As $(\bigcup_{I_0} \mathcal{A}_i)^{\cap \cup}$ is an algebra and as $\mathfrak{J}_0$ is directed by the relation of inclusion, (0.1) yields that $\bigcup_{\mathfrak{J}_0} (\bigcup_{I_0} \mathcal{A}_i)^{\cap \cup}$ is an algebra. As it contains $\bigcup_I \mathcal{A}_i$, this concludes the proof.

Lemma (0.3) Let $(\mathcal{A}_i^*)_{i \in I}$ be a family of σ-algebras on X and $\mathcal{A}_0 \subset \alpha(\bigcup_I \mathcal{A}_i^*)$ $[\mathcal{A}_0^* \subset \sigma(\bigcup_I \mathcal{A}_i^*)]$ be a countable subalgebra [separable sub-σ-algebra]. Then there exist countable systems $\mathcal{S}_i \subset \mathcal{A}_i^*$ such that $\mathcal{A}_0 \subset \alpha(\bigcup_I \mathcal{S}_i)$ $[\mathcal{A}_0^* \subset \sigma(\bigcup_I \mathcal{S}_i)]$.

<u>Proof:</u> Let $\{A_1, A_2, \ldots\}$ with $A_n \in \alpha(\bigcup_I \mathcal{A}_i^*)$ $[A_n \in \sigma(\bigcup_I \mathcal{A}_i^*)]$ $n = 1,2,\ldots$ be a countable sub-algebra $\mathcal{A}_0$ [countable system generating the separable sub-σ-algebra $\mathcal{A}_0^*$]. To each A_n there exists a finite [countable] system of sets $\mathcal{T}_n \subset \bigcup_I \mathcal{A}_i^*$ such that $A_n \in \alpha(\mathcal{T}_n)$ $[A_n \in \sigma(\mathcal{T}_n)]$. Let $\mathcal{S}_i := \mathcal{A}_i^* \cap (\bigcup_1^\infty \mathcal{T}_n)$. As $\bigcup_1^\infty \mathcal{T}_n$ is countable, $\mathcal{S}_i$ is countable for each $i \in I$. We have $\mathcal{T}_n \subset \bigcup_I \mathcal{S}_i$ and therefore $\alpha(\mathcal{T}_n) \subset \alpha(\bigcup_I \mathcal{S}_i)$ $[\sigma(\mathcal{T}_n) \subset \sigma(\bigcup_I \mathcal{S}_i)]$ for all $n = 1,2,\ldots$, whence $\mathcal{A}_0 \subset \alpha(\bigcup_I \mathcal{S}_i)$ $[\mathcal{A}_0^* \subset \sigma(\bigcup_I \mathcal{S}_i)]$ follows.

<u>Lemma (0.4)</u> Let $(\mathcal{A}_i^*)_{i \in I}$ be a family of σ-algebras on X and let $\mathcal{A}_0 \subset \bigcup_{\mathfrak{I}_0} \sigma(\bigcup_{I_0} \mathcal{A}_i^*)$ be a countable sub-algebra. Then there exist countable systems $\mathcal{S}_i \subset \mathcal{A}_i^*$ such that $\mathcal{A}_0 \subset \bigcup_{\mathfrak{I}_0} \sigma(\bigcup_{I_0} \mathcal{S}_i)$.

<u>Proof:</u> Let $\mathcal{A}_0 = \{A_1, A_2, \ldots\}$. To each n there exists $I_n \in \mathfrak{I}_0$ such that $A_n \in \sigma(\bigcup_{I_n} \mathcal{A}_i^*)$. Hence to each n there exists a countable system $\mathcal{T}_n \subset \bigcup_{I_n} \mathcal{A}_i^*$ such that $A_n \in \sigma(\mathcal{T}_n)$. Let $\mathcal{S}_i := \mathcal{A}_i^* \cap (\bigcup_1^\infty \mathcal{T}_n)$. Of course, $\mathcal{S}_i$ is countable for each $i \in I$. As $\mathcal{T}_n \subset \bigcup_{I_n} \mathcal{A}_i^*$, we have $\mathcal{T}_n \subset \bigcup_{I_n} \mathcal{S}_i$. Therefore, $\sigma(\mathcal{T}_n) \subset \bigcup_{\mathfrak{I}_0} \sigma(\bigcup_{I_0} \mathcal{S}_i)$ for $n = 1,2,\ldots$, whence $\mathcal{A}_0 \subset \bigcup_{\mathfrak{I}_0} \sigma(\bigcup_{I_0} \mathcal{S}_i)$ follows.

<u>Lemma (0.5)</u> A separable σ-algebra $\mathcal{A}^*$ on X can be induced by a function, i.e. there exists $f: X \to R^1$ such that $\mathcal{A}^* = f^{-1}\mathcal{B}^1$ (where $\mathcal{B}^1$ is the Borel - algebra of R^1).

<u>Proof:</u> Let $\mathcal{A}^* = \sigma(\{A_1, A_2, \dots\})$. We define the characteristic function of $(A_n)_{n=1,2,\dots}$ by

$$f(x) := \sum_{1}^{\infty} \frac{1}{3^n} \chi_{A_n}(x).$$

As f is $\mathcal{A}^*, \mathcal{B}^1$ - measurable, $f^{-1}\mathcal{B}^1 \subset \mathcal{A}^*$. The proof is concluded if we show $A_n \in f^{-1}\mathcal{B}^1$ for $n=1,2,\dots$. Let

$$B_n := \bigcup_{\substack{1 \le i \le n-1 \\ 1 \le r_1 < \dots < r_i \le n-1}} \left[\sum_{1}^{i} \frac{1}{3^{r_\nu}} + \frac{1}{3^n}, \sum_{1}^{i} \frac{1}{3^{r_\nu}} + \frac{1}{2 \cdot 3^{n-1}} \right] \cup \left[\frac{1}{3^n}, \frac{1}{2 \cdot 3^{n-1}} \right].$$

We have $B_n \in \mathcal{B}^1$. Furthermore, for each $x \in A_n$ we have $f(x) \in B_n$ and for $x \in A_n^c$ we have $f(x) \in B_n^c$. Hence $A_n = f^{-1} B_n$.

Notations

$a: = b$	a is defined as b
T/D	a map with domain of definition D
$T: X \to Y$	a map T/X into Y
$\emptyset$	the empty set
X	the basic abstract space
$A, B, C, \ldots$	subsets of X
A^c	the set X - A
$\chi_A(x)$	the characteristic function of the set A
$\mathcal{A}, \mathcal{B}, \mathcal{C}, \mathcal{S}$	systems of subsets of X
$\mathfrak{P}(X)$	the system of all subsets of X
$+$ or Σ	union of disjoint sets
$\mathcal{A}^c$	the system of sets A^c with $A \in \mathcal{A}$
$\mathcal{A}^{\cup}$ $(\mathcal{A}^{\cap})$	the smallest system of sets containing $\mathcal{A}$ which is closed under finite unions (intersections)
$\mathcal{A}^+$	the smallest system of sets containing $\mathcal{A}$ which is closed under finite disjoint unions
$\mathcal{A}^{\sigma}$ $(\mathcal{A}^{\delta})$	the smallest system of sets containing $\mathcal{A}$ which is closed under countable unions (intersections)
$\alpha(\mathcal{A})$ $(\sigma(\mathcal{A}))$	the smallest algebra (σ-algebra) on X containing $\mathcal{A}$
$(X, \mathcal{A})$ measurable space	$\mathcal{A}$ is a σ-algebra of subsets of X
T is $\mathcal{A}, \mathcal{B}$ - measurable	$T^{-1}\mathcal{B} \subset \mathcal{A}$

content	a non-negative and additive set function, defined on an algebra and vanishing at $\emptyset$
p-content	probability content, i.e. a content which takes the value 1 for the whole basic space
measure	a σ-additive content defined on a σ-algebra
p-measure	probability measure

References

Doob, J.L.

Stochastic processes, New York 1952.

Gnedenko, B.V., and A.N. Kolmogorov

Limit distributions for sums of independent random variables, (in Russian) Moscow - Leningrad 1949. English translation: Cambridge (Mass.) 1954.+)

Halmos, P.R.

Measure theory, Princeton (New Jersey) 1962.

Hewitt, E., and K.A. Ross

Abstract harmonic analysis I, Berlin - Göttingen - Heidelberg 1963.

Jiřina, M.

Conditional probabilities on σ-algebras with countable basis, (in Russian) Čzechoslovak Math. Journal 4 (79) (1954), pp 372 - 380. English translation: Selected Translations in Math. Statistics and Probability, vol. 2 (1962), pp 79 - 86.+)

Kappos, D.A.

Strukturtheorie der Wahrscheinlichkeitsfelder und -räume, Berlin - Göttingen - Heidelberg 1960.

Kelley, J.L.

General topology, New York 1963

Kolmogorov, A.N.

Grundbegriffe der Wahrscheinlichkeitsrechnung, Berlin 1933

+) Quotations are given for the English translations

Loève, M.

Probability theory, Princeton (New Jersey) 1963

Marczewski, E.

On compact measures, Fundamenta Mathematicae, vol. 40 (1953), pp 113 - 124

Ryll - Nardzewski, C.

On quasi - compact measures, Fundamenta Mathematicae, vol. 40 (1953), pp 125 - 130

Sierpinski, W.

General topology, Toronto 1952